Two Grade Arithmetic

BOOK TWO

By K. LOVELL, B.Sc., M.A., Ph.D.(Lond.)

Professor of Educational Psychology, University of Leeds

and C. H. J. SMITH, B.Sc., M.A., Ph.D.(Lond.)

formerly Senior Lecturer in Education and Methodology of Mathematics, Borough Road College

GINN AND COMPANY LTD
LONDON AND AYLESBURY

407604

Revised and reset 1969
First metric edition 1971
Revised metric edition 1975
Second impression 1976

Published by Ginn and Company Ltd.
Elsinore House, Buckingham Street, Aylesbury, Bucks HP20 2NQ

Product No. 112110339 ISBN 0 602 21883 7 (Pupils)

Product No. 112110428 ISBN 0 602 21887 X (Teachers)

Printed in Great Britain at the University Press, Oxford
by Vivian Ridler, Printer to the University

PREFACE

I hope that these books will give you experience of working some important kinds of examples. Those on the right hand page are usually a little harder than those on the left.

Your teacher will tell you which examples you should work. If you try hard and work steadily and neatly, I hope you will get right all the exercises that you attempt, and that you come to enjoy mathematics.

Leeds
1971

K. LOVELL

ADDITION

First try these

1 32 +24	**2** 15 +62	**3** 64 +17	**4** 40 +29
5 44 +15	**6** 51 +19	**7** 72 +18	**8** 87 +8
9 57 +16	**10** 37 +15	**11** 128 +342	**12** 195 +273
13 205 +495	**14** 338 +127	**15** 617 +95	**16** 827 +246
17 235 +987	**18** 694 +865	**19** 732 +309	**20** 495 +685
21 487 265 +108	**22** 329 174 +682	**23** 288 832 +145	**24** 589 146 +203
25 3472 +2103	**26** 4283 +3611	**27** 4017 +1520	**28** 6502 +1360
29 2197 +4028	**30** 4135 +1926	**31** 1485 +3951	**32** 3682 +2199
33 3715 +2988	**34** 2567 +6814	**35** 3849 +1782	**36** 6121 +3079

ADDITION

Now try these

1 56 +25	**2** 78 +17	**3** 47 +36	**4** 88 +13
5 20 96 +29	**6** 178 193 +456	**7** 258 167 +196	**8** 388 452 +134
9 462 179 +344	**10** 658 476 +392	**11** 3876 +1253	**12** 5417 +2496
13 2068 +3145	**14** 4982 +1208	**15** 6175 +1357	**16** 2191 +3058
17 1347 7205 +316	**18** 6129 2805 +17	**19** 2133 1685 +3907	**20** 4040 3219 +1234
21 1927 1418 +3178	**22** 4892 16 +1537	**23** 1062 3409 +2876	**24** 3297 1459 +2580
25 3062 2151 +538	**26** 2109 1280 +4175	**27** 1521 3729 +2108	**28** 2375 2168 +809
29 1950 6516 +98	**30** 1723 6581 +25	**31** 4965 1147 +2008	**32** 1883 1496 +5203

SUBTRACTION

First try these

1	36 - 12	**2**	27 - 13	**3**	35 - 25	**4**	49 - 36
5	68 - 24	**6**	71 - 40	**7**	84 - 51	**8**	68 - 55
9	73 - 60	**10**	94 - 33	**11**	41 - 29	**12**	52 - 46
13	75 - 38	**14**	82 - 47	**15**	93 - 59	**16**	437 - 215
17	641 - 418	**18**	715 - 483	**19**	672 - 349	**20**	825 - 179
21	363 - 274	**22**	500 - 217	**23**	908 - 689	**24**	3215 - 1703
25	2491 - 1265	**26**	7532 - 4419	**27**	2075 - 1058	**28**	8631 - 2951
29	7629 - 3847	**30**	5124 - 1251	**31**	4638 - 2559	**32**	9247 - 6573
33	4785 - 3887	**34**	6135 - 2478	**35**	5032 - 1486	**36**	4476 - 2987

SUBTRACTION

Now try these

1	62 - 35	**2**	81 - 53	**3**	94 - 28	**4**	72 - 39
5	471 - 258	**6**	582 - 137	**7**	418 - 162	**8**	600 - 153
9	4172 - 3421	**10**	8238 - 5174	**11**	9853 - 4219	**12**	6504 - 2160
13	4298 - 1743	**14**	8532 - 4175	**15**	7640 - 2950	**16**	5276 - 1384
17	6735 - 2198	**18**	5003 - 2161	**19**	4236 - 1387	**20**	8450 - 2693
21	6050 - 2357	**22**	7163 - 5784	**23**	8611 - 1876	**24**	3721 - 2954
25	6534 - 2755	**26**	8000 - 4183	**27**	4002 - 1239	**28**	3542 - 2693
29	4640 - 1835	**30**	6001 - 1007	**31**	7058 - 2989	**32**	9320 - 1479
33	8214 - 3757	**34**	4032 - 1864	**35**	3176 - 1277	**36**	5200 - 3481

MULTIPLICATION

First try these

1	4 ×3	**2**	5 ×4	**3**	5 ×6	**4**	8 ×4
5	7 ×6	**6**	8 ×3	**7**	6 ×5	**8**	7 ×5
9	9 ×3	**10**	9 ×4	**11**	12 ×4	**12**	13 ×2
13	14 ×5	**14**	12 ×6	**15**	15 ×3	**16**	16 ×4
17	17 ×4	**18**	18 ×5	**19**	19 ×3	**20**	10 ×4
21	20 ×3	**22**	30 ×3	**23**	40 ×4	**24**	23 ×3
25	24 ×4	**26**	27 ×3	**27**	34 ×6	**28**	143 ×7
29	165 ×8	**30**	173 ×9	**31**	241 ×7	**32**	238 ×7
33	316 ×8	**34**	336 ×8	**35**	425 ×9	**36**	417 ×9

MULTIPLICATION

Now try these

1 6×4	**2** 7×4	**3** 8×6	**4** 7×7
5 8×8	**6** 8×7	**7** 9×6	**8** 9×8
9 24×2	**10** 32×3	**11** 13×4	**12** 14×6
13 13×6	**14** 23×5	**15** 30×2	**16** 40×3
17 35×3	**18** 44×4	**19** 52×4	**20** 124×3
21 136×3	**22** 127×3	**23** 162×3	**24** 171×4
25 263×4	**26** 271×4	**27** 532×9	**28** 475×12
29 638×9	**30** 298×11	**31** 613×9	**32** 646×9
33 714×11	**34** 814×12	**35** 825×12	**36** 847×10

DIVISION

First try these

Set 1

1 3)15	**2** 3)18	**3** 3)24	**4** 3)42
5 3)72	**6** 4)16	**7** 4)24	**8** 4)56
9 4)96	**10** 4)52	**11** 3)28	**12** 3)45
13 3)69	**14** 3)105	**15** 3)127	**16** 4)40
17 4)60	**18** 4)78	**19** 4)104	**20** 4)135
21 5)35	**22** 5)60	**23** 5)95	**24** 5)143
25 5)189	**26** 6)54	**27** 6)73	**28** 6)99
29 6)156	**30** 6)183	**31** 3)369	**32** 3)393

Set 2

1 $159 \div 3$	**2** $186 \div 3$	**3** $204 \div 4$	**4** $484 \div 4$
5 $248 \div 4$	**6** $327 \div 4$	**7** $156 \div 5$	**8** $255 \div 5$
9 $315 \div 5$	**10** $603 \div 5$	**11** $126 \div 6$	**12** $620 \div 6$
13 $667 \div 6$	**14** $849 \div 6$	**15** $133 \div 7$	**16** $287 \div 7$
17 $429 \div 7$	**18** $567 \div 7$	**19** $484 \div 8$	**20** $649 \div 8$
21 $728 \div 8$	**22** $819 \div 9$	**23** $911 \div 9$	**24** $599 \div 9$
25 $658 \div 9$	**26** $729 \div 9$	**27** $3624 \div 3$	**28** $2824 \div 4$
29 $3015 \div 5$	**30** $3630 \div 6$	**31** $2821 \div 7$	**32** $2432 \div 8$

DIVISION

Now try these

Set 1

1	4)48	**2**	4)88	**3**	4)70	**4**	4)94
5	4)148	**6**	5)40	**7**	5)69	**8**	5)85
9	5)105	**10**	5)163	**11**	6)48	**12**	6)79
13	6)172	**14**	6)275	**15**	6)402	**16**	7)56
17	7)147	**18**	7)245	**19**	7)364	**20**	7)581
21	8)960	**22**	8)804	**23**	8)795	**24**	9)83
25	9)468	**26**	5)775	**27**	6)920	**28**	7)951
29	8)912	**30**	9)1098	**31**	9)4680	**32**	12)2532

Set 2

1	$693 \div 9$	**2**	$372 \div 10$	**3**	$1212 \div 12$	**4**	$647 \div 8$
5	$736 \div 8$	**6**	$968 \div 8$	**7**	$569 \div 10$	**8**	$639 \div 9$
9	$828 \div 9$	**10**	$909 \div 9$	**11**	$1089 \div 9$	**12**	$803 \div 11$
13	$608 \div 11$	**14**	$7766 \div 11$	**15**	$562 \div 11$	**16**	$1095 \div 12$
17	$984 \div 12$	**18**	$876 \div 12$	**19**	$1274 \div 12$	**20**	$4832 \div 8$
21	$8163 \div 9$	**22**	$4634 \div 11$	**23**	$3649 \div 12$	**24**	$4249 \div 7$
25	$4864 \div 8$	**26**	$5632 \div 8$	**27**	$6372 \div 9$	**28**	$8180 \div 9$
29	$7263 \div 9$	**30**	$6480 \div 10$	**31**	$7426 \div 11$	**32**	$9686 \div 12$

PROBLEMS

First try these

1 Write in figures seven thousand and twenty-four.

2 In a large garden there are 18 beech trees and 24 birch trees. How many trees are there altogether?

3 On the shelves of a class library are 42 books, while 19 other books are on loan to the pupils. How many books are there in the library altogether?

4 A Boy Scout troop of 54 scouts is divided up into 6 patrols. How many scouts are there in a patrol?

5 In a train there are 273 second-class and 148 first-class passengers. How many is this altogether?

6 A boy is given a 12 m start in a race of 220 m. How far has he to run?

7 Our lawn is 42 m long and our neighbour's lawn is 71 m long. How much longer is their lawn than ours?

8 A boy takes 23 marbles from a bag containing 48. How many are left?

9 In a country school of 136 pupils, 29 are absent. How many are present?

10 A grocer bought 9 boxes each weighing 25 kg. How much sugar did he buy?

11 There are 24 hours in a day. How many hours are there in a week?

12 How many eggs are there in 8 score?

13 At a Sunday School party there were 224 children, and they sat four at a table. How many tables were needed?

14 My house has 16 windows with 8 panes of glass in each. How many panes of glass are there altogether?

15 A packet of writing paper for an examination contains 188 sheets. If each child is to be given 4 sheets, how many children can be supplied?

16 If 256 kg of sand is to be put into 8 bags of the same size, how much must be put into each?

PROBLEMS

Now try these

1. Write in figures nine thousand and ninety-eight.
2. Two hundred and twenty-seven boys and two hundred and thirty-one girls in a school have milk. How many is this altogether?
3. In June a school used 2643 bottles of milk and 1997 in May. How many more were used in June?
4. What number is left over when 288 is divided by 5?
5. How many minutes are there in 9 hours?
6. Find the total number of days in the months of January, February and March, in a leap year.
7. In my garden I have 14 rows of cabbages, with 8 in a row. How many cabbages are there altogether?
8. A school of 144 children is to be taken for an excursion in 4 coaches all of the same size. How many children will travel in each coach?
9. In a box of 9 dozen eggs, 8 are bad. How many good eggs are there?
10. Share 129 oranges equally among 9 boys. How many will each get, and how many are over?
11. In four innings a cricketer scored 113, 96, 75 and 102 runs. What was his total score?
12. On four days the following numbers of people visited a museum: 253, 146, 325, 248. How many visitors were there?
13. How many eggs are there in 25 dozen?
14. If 177 apples are divided equally among 12 girls, how many will each receive, and how many will be left over?
15. A clothes line is 9 m long. What length of rope will be needed to make 12 clothes lines of this length?
16. A shopkeeper sold 83 bars of chocolate from a packet containing 12 dozen. How many bars were left?
17. A farmer has 128 hens which are to be put into 8 coops. How many should be put into each?

SPEED TESTS

First try these		Now try these	
1	$8+9$	**1**	$19+24+17$
2	$11-4$	**2**	$48-16$
3	$14-6$	**3**	32×6
4	11×11	**4**	$25+11+16$
5	$108\div9$	**5**	$102\div6$
6	$96\div8$	**6**	$105\div5$
7	$7+6$	**7**	$253-127$
8	35×4	**8**	36×7
9	$125\div5$	**9**	$138\div6$
10	$32-12$	**10**	$146+375$
11	$11+6+1$	**11**	$91-69$
12	$96\div6$	**12**	64×8
13	$104\div8$	**13**	$31+24+18$
14	$43-19$	**14**	$231\div7$
15	$91\div7$	**15**	163×9
16	9×6	**16**	85×6
17	14×0	**17**	$217-89$
18	$4+7+3$	**18**	$47-17$
19	$9+9$	**19**	$424\div8$
20	33×8	**20**	$666\div9$
21	$32-17$	**21**	$98+3+25$
22	$154\div7$	**22**	$594\div9$
23	74×8	**23**	98×9
24	$6+7+3$	**24**	$43+104+7$

SPEED TESTS

All can try these

Set 1

Put the correct sign in these exercises:

(*a*)	2	3	= 6	(*b*)	8	2	= 4
(*c*)	19	21	= 40	(*d*)	23	11	= 12
(*e*)	8	0	= 8	(*f*)	7	1	= 7
(*g*)	9	9	= 0	(*h*)	9	9	= 1
(*i*)	4	0	= 0	(*j*)	12	4	= 3
(*k*)	5	5	= 25	(*l*)	17	0	= 17
(*m*)	14	14	= 1	(*n*)	21	9	= 30

Set 2

The number 16 is equal to:

$$12 + 4$$
$$19 - 3$$
$$8 \times 2$$
$$32 \div 2$$

You will notice that four different signs have been used. Now write each of the following numbers in four ways:

1	12	**2**	20
3	8	**4**	14
5	9	**6**	18
7	24	**8**	10
9	3	**10**	32
11	21	**12**	15
13	6	**14**	19
15	17	**16**	28

LONG MULTIPLICATION (1)

First try these

1 16×10	**2** 18×10	**3** 24×10	**4** 31×10
5 18×13	**6** 13×13	**7** 20×13	**8** 16×13
9 24×14	**10** 19×14	**11** 27×14	**12** 32×14
13 13×15	**14** 17×15	**15** 24×15	**16** 28×15
17 16×16	**18** 23×16	**19** 29×16	**20** 20×16
21 21×17	**22** 30×17	**23** 25×17	**24** 18×17
25 24×18	**26** 18×18	**27** 31×18	**28** 22×18
29 25×19	**30** 21×19	**31** 32×19	**32** 46×19
33 37×16	**34** 53×17	**35** 45×19	**36** 28×18

LONG MULTIPLICATION (1)

Now try these

1 23 ×10	**2** 32 ×10	**3** 41 ×10	**4** 26 ×10
5 24 ×16	**6** 27 ×18	**7** 34 ×16	**8** 28 ×13
9 29 ×14	**10** 33 ×13	**11** 37 ×14	**12** 40 ×16
13 44 ×17	**14** 36 ×18	**15** 45 ×13	**16** 23 ×15
17 34 ×15	**18** 31 ×15	**19** 32 ×16	**20** 19 ×19
21 23 ×19	**22** 55 ×18	**23** 82 ×13	**24** 73 ×14
25 39 ×16	**26** 47 ×17	**27** 124 ×16	**28** 136 ×17
29 227 ×18	**30** 134 ×16	**31** 220 ×13	**32** 129 ×14
33 238 ×23	**34** 267 ×24	**35** 142 ×25	**36** 145 ×29

LONG DIVISION (1)

First try these

Set 1

1 $3\overline{)9}$	**2** $3\overline{)12}$	**3** $4\overline{)13}$	**4** $4\overline{)18}$
5 $6\overline{)27}$	**6** $6\overline{)32}$	**7** $4\overline{)84}$	**8** $4\overline{)87}$
9 $4\overline{)80}$	**10** $4\overline{)93}$	**11** $21\overline{)44}$	**12** $21\overline{)66}$
13 $21\overline{)70}$	**14** $31\overline{)66}$	**15** $31\overline{)95}$	**16** $22\overline{)45}$
17 $22\overline{)68}$	**18** $32\overline{)67}$	**19** $23\overline{)46}$	**20** $41\overline{)43}$
21 $41\overline{)88}$	**22** $41\overline{)97}$	**23** $31\overline{)70}$	**24** $41\overline{)90}$
25 $21\overline{)445}$	**26** $21\overline{)657}$	**27** $21\overline{)662}$	**28** $23\overline{)489}$
29 $31\overline{)651}$	**30** $32\overline{)672}$	**31** $41\overline{)863}$	**32** $42\overline{)886}$

Set 2

1 $21\overline{)43}$	**2** $21\overline{)49}$	**3** $21\overline{)65}$	**4** $21\overline{)86}$
5 $31\overline{)72}$	**6** $31\overline{)100}$	**7** $40\overline{)87}$	**8** $41\overline{)98}$
9 $32\overline{)68}$	**10** $32\overline{)99}$	**11** $42\overline{)95}$	**12** $42\overline{)89}$
13 $51\overline{)58}$	**14** $50\overline{)103}$	**15** $22\overline{)48}$	**16** $22\overline{)67}$
17 $23\overline{)48}$	**18** $23\overline{)70}$	**19** $20\overline{)51}$	**20** $23\overline{)69}$
21 $33\overline{)68}$	**22** $33\overline{)100}$	**23** $42\overline{)84}$	**24** $43\overline{)90}$
25 $21\overline{)443}$	**26** $22\overline{)487}$	**27** $21\overline{)656}$	**28** $31\overline{)659}$
29 $32\overline{)719}$	**30** $31\overline{)668}$	**31** $41\overline{)868}$	**32** $40\overline{)904}$

LONG DIVISION (1)

Now try these

Set 1

1 $21\overline{)53}$ **2** $21\overline{)92}$ **3** $21\overline{)85}$ **4** $21\overline{)46}$

5 $31\overline{)63}$ **6** $32\overline{)65}$ **7** $31\overline{)94}$ **8** $41\overline{)84}$

9 $41\overline{)91}$ **10** $23\overline{)54}$ **11** $23\overline{)75}$ **12** $32\overline{)70}$

13 $32\overline{)98}$ **14** $21\overline{)52}$ **15** $32\overline{)79}$ **16** $21\overline{)453}$

17 $21\overline{)652}$ **18** $20\overline{)674}$ **19** $24\overline{)744}$ **20** $25\overline{)750}$

21 $24\overline{)486}$ **22** $33\overline{)726}$ **23** $31\overline{)666}$ **24** $31\overline{)967}$

25 $32\overline{)678}$ **26** $32\overline{)999}$ **27** $44\overline{)493}$ **28** $40\overline{)894}$

29 $42\overline{)1280}$ **30** $43\overline{)479}$ **31** $54\overline{)1094}$ **32** $57\overline{)1786}$

Set 2

1 $21\overline{)147}$ **2** $21\overline{)196}$ **3** $20\overline{)105}$ **4** $21\overline{)126}$

5 $31\overline{)159}$ **6** $32\overline{)128}$ **7** $31\overline{)373}$ **8** $31\overline{)299}$

9 $41\overline{)47}$ **10** $40\overline{)85}$ **11** $43\overline{)89}$ **12** $42\overline{)130}$

13 $50\overline{)68}$ **14** $53\overline{)107}$ **15** $65\overline{)67}$ **16** $61\overline{)123}$

17 $21\overline{)449}$ **18** $20\overline{)675}$ **19** $36\overline{)439}$ **20** $33\overline{)679}$

21 $43\overline{)869}$ **22** $45\overline{)496}$ **23** $28\overline{)589}$ **24** $28\overline{)900}$

25 $29\overline{)320}$ **26** $29\overline{)611}$ **27** $38\overline{)799}$ **28** $38\overline{)815}$

29 $39\overline{)820}$ **30** $39\overline{)430}$ **31** $48\overline{)490}$ **32** $48\overline{)1000}$

LONG MULTIPLICATION (2)

First try these

1 24×22	**2** 35×22	**3** 43×25	**4** 36×20
5 34×23	**6** 46×29	**7** 29×34	**8** 53×27
9 62×33	**10** 57×30	**11** 33×32	**12** 49×24
13 48×26	**14** 63×25	**15** 43×28	**16** 47×29
17 37×34	**18** 54×35	**19** 64×36	**20** 51×37
21 56×44	**22** 57×39	**23** 70×42	**24** 61×41
25 58×37	**26** 59×50	**27** 62×58	**28** 71×49
29 68×32	**30** 72×53	**31** 73×51	**32** 64×59
33 65×47	**34** 67×53	**35** 72×46	**36** 80×65

LONG MULTIPLICATION (2)

Now try these

1	37×21	**2**	45×24	**3**	56×26	**4**	68×27
5	38×34	**6**	43×37	**7**	64×39	**8**	67×45
9	63×47	**10**	59×48	**11**	73×52	**12**	84×56
13	93×57	**14**	92×50	**15**	77×65	**16**	83×67
17	82×68	**18**	83×71	**19**	80×73	**20**	92×76
21	79×78	**22**	86×82	**23**	94×83	**24**	79×85
25	90×89	**26**	95×91	**27**	99×97	**28**	68×83
29	69×48	**30**	74×63	**31**	86×69	**32**	96×86
33	94×80	**34**	89×78	**35**	84×77	**36**	92×87

LONG DIVISION (2)

First try these

Set 1

1 $29\overline{)31}$	**2** $29\overline{)61}$	**3** $29\overline{)87}$	**4** $28\overline{)57}$
5 $28\overline{)85}$	**6** $28\overline{)90}$	**7** $27\overline{)83}$	**8** $27\overline{)56}$
9 $27\overline{)86}$	**10** $39\overline{)79}$	**11** $39\overline{)81}$	**12** $39\overline{)85}$
13 $38\overline{)77}$	**14** $36\overline{)72}$	**15** $38\overline{)87}$	**16** $37\overline{)75}$
17 $37\overline{)41}$	**18** $37\overline{)81}$	**19** $26\overline{)54}$	**20** $26\overline{)81}$
21 $26\overline{)89}$	**22** $36\overline{)75}$	**23** $25\overline{)76}$	**24** $29\overline{)610}$
25 $39\overline{)429}$	**26** $28\overline{)871}$	**27** $27\overline{)839}$	**28** $37\overline{)783}$
29 $38\overline{)419}$	**30** $36\overline{)398}$	**31** $49\overline{)1029}$	**32** $49\overline{)1519}$

Set 2

1 $29\overline{)65}$	**2** $28\overline{)88}$	**3** $29\overline{)59}$	**4** $28\overline{)64}$
5 $39\overline{)82}$	**6** $39\overline{)48}$	**7** $27\overline{)92}$	**8** $27\overline{)63}$
9 $38\overline{)80}$	**10** $38\overline{)89}$	**11** $24\overline{)81}$	**12** $24\overline{)56}$
13 $25\overline{)84}$	**14** $25\overline{)63}$	**15** $26\overline{)83}$	**16** $27\overline{)89}$
17 $34\overline{)69}$	**18** $36\overline{)78}$	**19** $35\overline{)73}$	**20** $37\overline{)76}$
21 $46\overline{)94}$	**22** $29\overline{)609}$	**23** $28\overline{)589}$	**24** $28\overline{)869}$
25 $27\overline{)838}$	**26** $29\overline{)640}$	**27** $27\overline{)864}$	**28** $37\overline{)779}$
29 $37\overline{)816}$	**30** $38\overline{)799}$	**31** $35\overline{)1785}$	**32** $35\overline{)2486}$

LONG DIVISION (2)

Now try these

Set 1

1 $49\overline{)494}$	**2** $48\overline{)529}$	**3** $48\overline{)578}$	**4** $49\overline{)580}$
5 $47\overline{)989}$	**6** $47\overline{)518}$	**7** $47\overline{)1039}$	**8** $46\overline{)508}$
9 $46\overline{)967}$	**10** $46\overline{)1035}$	**11** $45\overline{)496}$	**12** $45\overline{)999}$
13 $45\overline{)947}$	**14** $57\overline{)1824}$	**15** $57\overline{)2793}$	**16** $58\overline{)3422}$
17 $58\overline{)4872}$	**18** $59\overline{)5192}$	**19** $59\overline{)5546}$	**20** $61\overline{)3538}$
21 $61\overline{)4514}$	**22** $63\overline{)4347}$	**23** $63\overline{)5733}$	**24** $67\overline{)4087}$
25 $67\overline{)5092}$	**26** $69\overline{)5244}$	**27** $69\overline{)6279}$	**28** $71\overline{)4473}$
29 $71\overline{)5893}$	**30** $73\overline{)4454}$	**31** $75\overline{)5327}$	**32** $77\overline{)5930}$

Set 2

1 $35\overline{)2498}$	**2** $35\overline{)3045}$	**3** $36\overline{)1695}$	**4** $36\overline{)2376}$
5 $36\overline{)3315}$	**6** $37\overline{)2184}$	**7** $37\overline{)2701}$	**8** $37\overline{)3036}$
9 $45\overline{)2790}$	**10** $45\overline{)3780}$	**11** $46\overline{)3544}$	**12** $46\overline{)4370}$
13 $47\overline{)2399}$	**14** $47\overline{)4371}$	**15** $68\overline{)3468}$	**16** $68\overline{)5092}$
17 $68\overline{)6098}$	**18** $71\overline{)1919}$	**19** $71\overline{)3337}$	**20** $71\overline{)6108}$
21 $74\overline{)4144}$	**22** $74\overline{)5254}$	**23** $74\overline{)6734}$	**24** $78\overline{)2028}$
25 $78\overline{)3590}$	**26** $78\overline{)5382}$	**27** $78\overline{)7490}$	**28** $79\overline{)4819}$
29 $79\overline{)7031}$	**30** $79\overline{)7742}$	**31** $81\overline{)5751}$	**32** $83\overline{)7138}$

PROBLEMS

First try these

1 How far will a train travel in 14 hours if, allowing for stops, it can travel 72 km in each hour?

2 Share 65 marbles among 31 boys. What is each boy's share? How many are left over?

3 At a children's party 17 tables were arranged to seat 16 children at each. How many children were at the party?

4 Twenty-eight boxes each held 72 apples. How many apples was this altogether?

5 If 441 oranges are divided equally among 21 children, how many will each get?

6 If 360 cakes are divided equally among 15 tables, how many will be placed on each table?

7 In a school there are 18 classes and 38 children in each. How many children are there on roll?

8 Share 675 cards between 32 girls. How many will each get, and how many will be left over?

9 If a bus holds 54 people, how many people can be carried by 24 buses?

10 A girl took 19 minutes to walk to school. She took 98 steps a minute. How many steps did she take in walking to school?

11 An article costs £25·00. How many similar articles can be bought for £580·00, and how much money will be left over?

12 What will a man earn in 42 weeks if he earns £39·00 per week?

13 Divide 806 by 26, and add 10 to your answer.

14 How many times can 46 be subtracted from 966?

15 Divide 324 by 18, and multiply your result by 15.

16 One farmer had 33 sheep and another had 22 times as many. How many more sheep had the second farmer than the first?

17 A country postman cycles 96 km every week. How many km will he cycle in 39 weeks?

PROBLEMS

Now try these

1 A motor coach holds 38 children. If 456 children were taken on a day's outing, how many coaches were needed?
2 Biscuits are packed into boxes, each of which holds 60. How many biscuits can be packed in $2\frac{1}{2}$ dozen boxes?
3 How many times can 44 be taken from 946? How many over?
4 In 49 hours a liner steamed 1323 sea miles. How many sea miles an hour is this?
5 Fifty boxes each hold four dozen pencils. How many pencils is this altogether?
6 If a school has 396 pairs of plimsolls, how many pairs can be given to each of 18 classes?
7 Multiply 98 by 61, and take 100 from the result.
8 A boiler used 85 kg of coal per week. How much coal does it use in 1 year (52 weeks)?
9 What is the result of dividing 1276 by 29?
10 How many lengths of string, each 43 cm long, can be cut from a ball of string 1080 cm long? How much will be left over?
11 Find the product of 79 and 65. Take the result from 10 000.
12 How often is 61 contained in the number 41 less than 1000?
13 The weight of an article is 14 kg. How many such articles will together weigh 3360 kg?
14 There are 24 hours in a day. How many days are there in 1250 hours?
15 How many matches are there in half a gross of boxes, each of which holds 48 matches?
16 What is the nearest number to 900 that will divide exactly by 37?
17 Divide the product of 35 and 27 by 17.
18 How many score are there in 1428, and what is the remainder?

GENERAL REVISION

First try these

Add

1	42	**2**	225	**3**	245	**4**	2493
	273		714		1577		140
	+334		+146		+36		+623

Subtract

5	472	**6**	527	**7**	5805	**8**	6556
	−239		−294		−2490		−4194

Multiply

9 51×24 **10** 49×43 **11** 65×37 **12** 71×58

Divide

13 $559 \div 42$ **14** $756 \div 34$ **15** $625 \div 51$ **16** $626 \div 29$

17 Write in figures six thousand eight hundred and ten.

18 Make the largest number you can from the figures 3, 4, 2, and multiply it by 9.

19 The customers of a toy shop during the four weeks of November were: 123, 155, 197, 226. How many customers visited the shop during this time?

20 How many times can 21 be taken from 89, and what is the remainder?

21 A boy wrote down thirteen nineteens and added them up. Find the answer by a quicker method.

22 A girl took 103 from a certain number and her answer was 437. What was the number?

23 In four days a bus carried 342, 156, 128 and 317 people to the station. How many people short of 1000 did it carry in these four days?

GENERAL REVISION

Now try these

Add

1	563	**2**	475	**3**	3594	**4**	693
	417		1914		2571		214
	1524		326		464		3377
	+218		+619		+397		+2685

Subtract

5	617	**6**	704	**7**	7859	**8**	9768
	-263		-258		-6399		-570

Multiply

9 63×28 **10** 86×59 **11** 76×47 **12** 92×85

Divide

13 $6341 \div 29$ **14** $7341 \div 37$ **15** $8473 \div 56$ **16** $7576 \div 65$

17 Write in figures eight thousand and nineteen.

18 Take the smallest number you can make from the figures 3, 7, 2, 1, from the largest number you can make with them.

19 The number of people visiting a fair on four days were: 3421, 897, 2468, 1992. How many people attended the fair?

20 Share 165 marbles equally among 41 boys. How many does each get, and how many are left over?

21 We have three bottles of milk every day. How many do we have altogether in one ordinary year, and how many in one leap year?

22 A bus holds 56 people. How many people will a fleet of 25 buses hold?

23 How many times can 71 be subtracted from 895, and what will be the remainder?

24 Take one-quarter of 512 from one-half of 3952.

MONEY ADDITION

First try these

1 p 04 + 17

2 p 23 + 19

3 p 40 + 26

4 p 54 + 38

5 p 20 + 15

6 p 57 + 34

7 p $49\frac{1}{2}$ + $41\frac{1}{2}$

8 p 68 + 21

9 £ 0·36 + 0·51

10 £ 2·83 + 3·17

11 £ $0{\cdot}71\frac{1}{2}$ + $4{\cdot}25\frac{1}{2}$

12 £ 5·13 + 6·07

13 £ 8·95 + 3·72

14 £ 10·01 + 4·89

15 £ 0·30 + 0·76

16 £ $7{\cdot}59\frac{1}{2}$ + $6{\cdot}43\frac{1}{2}$

17 £ 0·81 + 9·37

18 £ 6·19 + $11{\cdot}54\frac{1}{2}$

19 £ 5·28 + 7·62

20 £ 8·67 + 2·14

21 £ 13·96 + 19·05

22 £ 14·82 + 16·14

23 £ 15·35 + 11·21

24 £ 12·77 + $14{\cdot}30\frac{1}{2}$

25 £ $27{\cdot}61\frac{1}{2}$ + $12{\cdot}13\frac{1}{2}$

26 £ 24·09 + 65·48

27 £ 20·47 + 15·50

28 £ 4·20 + 19·02

MONEY ADDITION

Now try these

1 £ 3·27 +2·98	**2** £ 4·91½ +16·52	**3** £ 13·76½ +14·09½	**4** £ 22·15 +31·86½
5 £ 47·39 +29·64	**6** £ 38·60½ +103·21	**7** £ 15·42 +48·53½	**8** £ 25·07½ +42·66
9 £ 120·80½ +24·79½	**10** £ 64·57 +50·98	**11** £ 73·43 +2·05	**12** £ 35·56 +68·17
13 £ 32·96 14·14 +6·02	**14** £ 1·83 19·07 +25·60	**15** £ 22·16 31·50 +87·79	**16** £ 20·43 39·25 +0·37
17 £ 47·51½ 62·18½ +19·60½	**18** £ 34·18 51·70 +95·04½	**19** £ 65·39 74·46 +80·58	**20** £ 29·68½ 13·21 +71·40
21 £ 43·19 0·73½ +102·45	**22** £ 36·50 27·96 +15·87	**23** £ 90·01 47·63 +18·25½	**24** £ 56·29 14·70 +83·92

MONEY SUBTRACTION

First try these

1	p 16 $-$08	**2**	p 23 $-12\frac{1}{2}$	**3**	p 45 $-$29	**4**	p 80 $-$53
5	p 38 $-$15	**6**	p 44 $-$22	**7**	p 87 $-46\frac{1}{2}$	**8**	p $95\frac{1}{2}$ $-$64
9	£ 0·25 $-$0·09	**10**	£ 0·61 $-0{\cdot}58\frac{1}{2}$	**11**	£ 0·90 $-$0·37	**12**	£ 0·74 $-$0·45
13	£ 3·65 $-$1·20	**14**	£ 2·28 $-$1·07	**15**	£ 4·53 $-$2·42	**16**	£ 6·87 $-$5·64
17	£ 7·53 $-$3·46	**18**	£ 2·72 $-0{\cdot}38\frac{1}{2}$	**19**	£ 9·65 $-$6·19	**20**	£ 8·94 $-$8·07
21	£ 10·31 $-$7·24	**22**	£ 13·93 $-$6·54	**23**	£ 19·65 $-8{\cdot}76\frac{1}{2}$	**24**	£ 21·00 $-$4·63
25	£ 23·02 $-14{\cdot}82\frac{1}{2}$	**26**	£ 35·61 $-$19·68	**27**	£ 30·50 $-$0·74	**28**	£ 44·82 $-$26·56

MONEY SUBTRACTION

Now try these

1 £ 1·61 − 0·35	**2** £ 4·39 − 2·84½	**3** £ 10·00 − 1·52	**4** £ 13·72½ − 2·96
5 £ 18·97 − 15·68	**6** £ 27·81 − 19·94	**7** £ 41·73 − 34·85	**8** £ 37·61 − 25·79
9 £ 37·80 − 36·97	**10** £ 42·28 − 29·79	**11** £ 50·17½ − 21·52	**12** £ 69·61 − 38·63½
13 £ 40·09 − 17·43	**14** £ 75·13½ − 0·62½	**15** £ 63·04 − 29·61½	**16** £ 100·00 − 58·47
17 £ 135·21 − 132·84	**18** £ 155·01 − 97·76	**19** £ 262·80 − 140·53	**20** £ 141·34 − 16·69
21 £ 230·41 − 36·73	**22** £ 153·71 − 149·82	**23** £ 194·56 − 187·90½	**24** £ 270·60 − 45·12½
25 £ 201·50 − 106·87½	**26** £ 294·08 − 36·29	**27** £ 321·09 − 45·63	**28** £ 316·62½ − 110·50

MONEY MULTIPLICATION

First try these

1 p 12 ×7	**2** p 23½ ×4	**3** p 09 ×8	**4** p 05½ ×12
5 p 12 ×4	**6** p 08½ ×6	**7** p 21 ×3	**8** p 47½ ×2
9 £ 0·25 ×4	**10** £ 0·37 ×3	**11** £ 0·56½ ×2	**12** £ 0·41 ×5
13 £ 0·18 ×7	**14** £ 0·29 ×6	**15** £ 0·41 ×12	**16** £ 0·73 ×10
17 £ 1·14 ×7	**18** £ 3·25 ×3	**19** £ 2·16 ×6	**20** £ 4·08 ×11
21 £ 5·07½ ×12	**22** £ 4·25 ×8	**23** £ 2·38 ×9	**24** £ 6·81 ×10
25 £ 3·64 ×7	**26** £ 5·09½ ×12	**27** £ 6·17 ×11	**28** £ 5·72 ×9

MONEY MULTIPLICATION

Now try these

1 £ 4·36 ×8	**2** £ 5·47 ×8	**3** £ 10·78 ×8	**4** £ 11·82 ×9
5 £ 8·65 ×9	**6** £ 13·41 ×9	**7** £ 15·04 ×10	**8** £ 3·98 ×10
9 £ 12·37 ×10	**10** £ 7·15 ×11	**11** £ 5·82 ×11	**12** £ 14·13 ×12
13 £ 10·56 ×12	**14** £ 21·29 ×6	**15** £ 19·66 ×5	**16** £ 22·70 ×8
17 £ 16·81 ×12	**18** £ 23·09 ×7	**19** £ 12·93 ×11	**20** £ 25·37 ×10
21 £ 30·23 ×7	**22** £ 18·54 ×12	**23** £ 10·19½ ×8	**24** £ 27·48 ×11
25 £ 20·71½ ×9	**26** £ 17·06½ ×7	**27** £ 32·13 ×12	**28** £ 19·73½ ×8

MONEY DIVISION

First try these

1 p 4)32	**2** p 5)27½	**3** p 7)91	**4** p 6)57
5 p 11)38½	**6** p 8)72	**7** p 12)90	**8** p 9)54
9 £ 12)1·68	**10** £ 8)2·40	**11** £ 9)1·89	**12** £ 11)1·43
13 £ 4)8·64	**14** £ 3)9·27	**15** £ 5)12·60	**16** £ 7)29·61
17 £ 4)28·60	**18** £ 5)9·45	**19** £ 6)24·72	**20** £ 3)38·55
21 £ 7)37·10	**22** £ 4)60·54	**23** £ 5)49·32½	**24** £ 6)50·04
25 £ 3)62·97	**26** £ 4)59·12	**27** £ 6)65·58	**28** £ 5)99·15
29 £ 4)80·06	**30** £ 7)38·95½	**31** £ 8)42·16	**32** £ 6)75·63
33 £ 5)73·10	**34** £ 8)67·56	**35** £ 3)59·16	**36** £ 7)48·93

MONEY DIVISION

Now try these

1 £ 6)18·90	**2** £ 6)27·54	**3** £ 6)32·76	**4** £ 7)19·81
5 £ 7)25·48	**6** £ 7)18·83	**7** £ 8)21·60	**8** £ 8)33·28
9 £ 9)28·35	**10** £ 9)23·67	**11** £ 9)37·89	**12** £ 10)31·20
13 £ 10)42·50	**14** £ 10)53·90	**15** £ 11)48·73	**16** £ 11)61·49
17 £ 12)40·20	**18** £ 12)101·52	**19** £ 12)92·88	**20** £ 6)92·37
21 £ 6)110·96	**22** £ 7)103·81	**23** £ 7)127·05	**24** £ 8)119·24
25 £ 8)65·72	**26** £ 9)188·55	**27** £ 10)207·93	**28** £ 11)96·41½
29 £ 10)120·40	**30** £ 11)220·66	**31** £ 12)241·08	**32** £ 12)193·74
33 £ 9)152·69	**34** £ 8)200·38	**35** £ 12)104·04	**36** £ 11)238·15

MONEY PROBLEMS

First try these

1 A boy bought a cricket bat costing £5·60, a ball costing £2·25 and stumps costing £2·95. How much did he spend?

2 A man left £12·00 with his tailor, who was to make him a suit costing £47·00, and a sports coat costing £21·55. How much more money had he to pay?

3 Seven new balls for netball cost £29·68. How much were they each?

4 Barbara worked for 9 weeks, earning £21·25 each week. How much did she earn altogether?

5 At a sale a bicycle usually costing £33·50 was sold for £19·75. How much was saved by buying it during the sale?

6 If I share £5·00 equally among 7 boys, how much will each get, and how much will be left over?

7 At Betty's party the cakes cost £2·75, the fruit £1·12½, the jellies £0·48 and other things £3·85. Find the cost of the party.

8 Find the cost of a score of articles at £9·00 each.

9 Find the sum of £3·70, £2·76½ and £41·50, and take the result from £50·00.

10 A boy saves 12½p a week to buy National Savings Certificates, which cost £1·00 each. How many weeks will he have to save to buy (*a*) 1 certificate, and (*b*) 5 certificates?

11 How much does a man earn in 37 hours at 85p per hour?

12 Five articles cost £78·62½. What is the cost per article?

13 Take one-half of £25·00 from £32·17.

14 A man wishes to save £45·00 in 8 weeks. How much must he save each week?

15 How much must be added to £14·16 to make £23·84½?

16 Find the cost of 4 coffee tables at £5·57½ each.

MONEY PROBLEMS

Now try these

1 Find the cost of half a dozen children's cycles at £20·50 each.
2 Add together £52·30, £46·81 and £120·26.
3 How much is 10 times the sum of £5·00 and £0·12½?
4 A girl wishes to buy a present costing £7·15, but has only £2·16½. How much money must she save?
5 Mrs. Brown gave £5·00 to a shopkeeper who said, "Give me 2p and I will give you two pounds." How much did she spend?
6 In a sale £0·25 was taken off every £. How much shall I pay in the sale for an article which usually costs £22·50?
7 Share £43·12 equally among 6 girls and 5 boys.
8 A man and three boys earned £75·00. If the man earned £30·00, find how much each boy earned if all the boys received the same amount.
9 By how much is £150·07 greater than £92·85?
10 A dozen railway tickets of the same price cost £56·40. How much did each cost?
11 Take £32·41 from £68·05, and divide the result by 4.
12 A party of 9 children are going abroad with their teacher. The trip will cost each child £72·25. What will be the total cost to the 9 children?
13 A man owed £200·00. He repaid £40·20, £68·75 and £25·15. How much does he still owe?
14 Six similar dresses cost £95·40. What is the cost of each?
15 Add one-quarter of £104·50 to one-half of £133·86.
16 A clerk makes a sum of money come to £250·79; another makes the result £249·81. What is the difference in their results?
17 Divide £58·80 equally among 2 men, 3 women, 4 boys and 3 girls.

LENGTH ADDITION AND SUBTRACTION

(m and cm).

First try these

Set 1

	m	cm
1	3	20
	+4	31
2	5	16
	+9	43
3	4	35
		+27
4	6	3
	+10	82
5		67
	+8	9
6	2	56
	+7	39
7	4	37
	+9	60
8	10	82
	+4	28
9	8	68
	+1	59

Set 2

	m	cm
1	7	24
	−3	19
2	6	62
	−1	45
3	2	89
		−32
4	11	35
	−4	26
5	15	90
	−6	57
6	4	20
	−2	40
7	10	14
	−5	84
8	19	78
	−12	57
9	18	65
	−10	18
10	15	23
	−7	90
11	20	68
	−1	12
12	9	44
	−6	7

LENGTH ADDITION AND SUBTRACTION

(m and cm)

First try these

Set 1

1

	m	cm
	17	58
+	2	15

2

	m	cm
	29	65
+	14	43

3

	m	cm
	30	26
+	21	50

4

	m	cm
	25	62
+	41	38

5

	m	cm
	18	97
+	24	63

6

	m	cm
	49	85
+	3	9

7

	m	cm
	54	13
+	26	87

8

	m	cm
	32	48
+	93	29

9

	m	cm
	64	59
+	47	65

Set 2

1

	m	cm
	15	84
−	3	19

2

	m	cm
	22	51
−	17	65

3

	m	cm
	36	82
−	8	27

4

	m	cm
	28	50
−	12	20

5

	m	cm
	50	34
−	29	81

6

	m	cm
	43	26
−	35	68

7

	m	cm
	78	11
−	60	76

8

	m	cm
	81	19
−	35	40

9

	m	cm
	68	33
−	48	51

10

	m	cm
	103	95
−	82	16

11

	m	cm
	145	20
−	69	75

12

	m	cm
	200	0
−	109	28

LENGTH ADDITION AND SUBTRACTION

(*km and m*)

First try these

Set 1

1

km	m
3	200
+ 4	100

2

km	m
5	170
+ 8	90

3

km	m
7	250
+ 1	500

4

km	m
6	340
+ 9	280

5

km	m
	620
+ 11	85

6

km	m
2	430
+ 3	196

7

km	m
10	780
+	220

8

km	m
8	400
+ 6	750

9

km	m
1	382
+ 2	895

Set 2

1

km	m
7	590
− 2	315

2

km	m
5	250
− 1	65

3

km	m
10	700
− 3	400

4

km	m
12	910
− 8	560

5

km	m
18	695
− 13	87

6

km	m
23	116
− 21	12

7

km	m
19	178
− 5	144

8

km	m
23	980
− 6	230

9

km	m
20	521
− 3	83

10

km	m
16	285
− 3	610

11

km	m
29	107
− 14	491

12

km	m
22	360
− 12	127

LENGTH ADDITION AND SUBTRACTION

(km and m)

Now try these

Set 1

1

km	m
16	185
+ 1	51

2

km	m
23	74
+ 10	367

3

km	m
39	436
+ 27	504

4

km	m
28	742
+	813

5

km	m
56	931
+ 13	600

6

km	m
40	527
+ 19	498

7

km	m
69	82
+ 47	150

8

km	m
71	502
+ 10	833

9

km	m
96	463
+ 107	539

Set 2

1

km	m
6	531
− 5	688

2

km	m
11	900
− 7	422

3

km	m
23	297
− 15	149

4

km	m
16	81
− 2	43

5

km	m
29	468
− 13	295

6

km	m
43	29
− 18	814

7

km	m
93	716
− 48	329

8

km	m
56	302
− 21	991

9

km	m
98	574
− 69	208

10

km	m
173	100
− 24	536

11

km	m
300	230
− 199	427

12

km	m
251	425
− 183	160

LENGTH MULTIPLICATION

First try these

Set 1

1

m	cm
3	10
	×2

2

m	cm
4	8
	×2

3

m	cm
1	19
	×2

4

m	cm
5	25
	×2

5

m	cm
2	43
	×2

6

m	cm
7	50
	×2

7

m	cm
6	58
	×2

8

m	cm
9	70
	×2

9

m	cm
8	49
	×2

Set 2

1

km	m
1	34
	×2

2

km	m
3	50
	×2

3

km	m
5	160
	×2

4

km	m
6	425
	×2

5

km	m
4	370
	×2

6

km	m
2	500
	×2

7

km	m
8	69
	×2

8

km	m
7	614
	×2

9

km	m
3	700
	×2

10

km	m
10	81
	×2

11

km	m
6	395
	×2

12

km	m
5	810
	×2

LENGTH MULTIPLICATION

Now try these

Set 1

	m	cm			m	cm			m	cm
1	4	31		**2**	10	50		**3**	27	48
		×2				×2				×2
4	16	78		**5**	34	91		**6**	47	60
		×2				×2				×2
7	25	84		**8**	62	5		**9**	58	71
		×2				×2				×2

Set 2

	km	m			km	m			km	m
1	4	250		**2**	7	195		**3**	16	720
		×2				×2				×2
4	11	536		**5**	25	413		**6**	30	608
		×2				×2				×2
7	43	915		**8**	36	840		**9**	51	499
		×2				×2				×2
10	69	148		**11**	76	623		**12**	95	876
		×2				×2				×2

LENGTH DIVISION

First try these

Divide each of the following quantities by 2

Set 1

1 2 m 6 cm

2 10 m 14 cm

3 8 m 24 cm

4 6 m 30 cm

5 14 m 50 cm

6 4 m 28 cm

7 20 m 60 cm

8 18 m 54 cm

9 26 m 72 cm

10 16 m 88 cm

11 12 m 96 cm

12 28 m 62 cm

13 5 m 0 cm

14 11 m 14 cm

15 13 m 12 cm

16 7 m 4 cm

17 9 m 8 cm

18 15 m 20 cm

Set 2

1 4 km 100 m

2 10 km 40 m

3 6 km 220 m

4 8 km 600 m

5 16 km 440 m

6 14 km 520 m

7 28 km 680 m

8 22 km 900 m

9 34 km 750 m

10 38 km 398 m

11 40 km 830 m

12 36 km 452 m

13 3 km 0 m

14 9 km 100 m

15 11 km 200 m

16 7 km 80 m

17 13 km 240 m

18 1 km 400 m

LENGTH DIVISION

Now try these

Divide each of the following quantities by 2

Set 1

1 8 m 10 cm
2 32 m 40 cm
3 22 m 34 cm
4 17 m 26 cm
5 13 m 16 cm
6 21 m 38 cm
7 29 m 42 cm
8 50 m 0 cm
9 43 m 12 cm
10 47 m 60 cm
11 86 m 94 cm
12 65 m 48 cm
13 77 m 52 cm
14 82 m 98 cm
15 69 m 84 cm
16 95 m 30 cm
17 99 m 46 cm
18 71 m 94 cm

Set 2

1 10 km 80 m
2 24 km 100 m
3 30 km 240 m
4 19 km 200 m
5 26 km 300 m
6 37 km 10 m
7 31 km 420 m
8 49 km 360 m
9 48 km 178 m
10 56 km 790 m
11 43 km 690 m
12 51 km 526 m
13 59 km 236 m
14 64 km 812 m
15 75 km 904 m
16 85 km 790 m
17 67 km 380 m
18 93 km 938 m

LENGTH PROBLEMS

(m and cm)

First try these

Set 1

1 Take 48 cm from 5 m and divide the result by 2 (answer in cm).

2 Find the difference in height between a girl who measures 114 cm and her mother, who is 1 m 62 cm tall.

3 A man buys 3 pieces of wood measuring 10 m, 11 m 36 cm and 6 m 75 cm. What length of wood did he buy altogether?

4 What length of rope is required to make 2 clothes lines each 14 m 50 cm long?

5 I have some slabs of concrete each 500 cm long. If I lay 5 of them end to end to make a path, how long is the path in metres?

6 Our garage is 4 m 80 cm long and when out car is in it there is a space of 25 cm at each end. How long is the car?

Set 2

1 A man can reach to a height of 2 m 20 cm. He stands on top of a pair of steps which raise him by 2 m 1 cm. To what height can he reach now?

2 A boy's model railway has a track which measures 13 m 10 cm round. How far does the train travel in going round the track twice?

3 A piece of wire 7 m 24 cm long is cut into two equal lengths. How long was each piece?

4 What remains when 12 m 67 cm is taken from 54 m 38 cm?

5 Jack, Bill and Mary live in the same road. It is 95 m from Jack's gate to Bill's gate, and 283 m from Jack's gate to Mary's gate. How far is it from Mary's to Bill's?

6 The three sides of a triangle are 8 cm, 13 cm, and 19 cm long. Find one half of the distance around the triangle.

7 What must be added to 26 m to make the result equal to difference between 50 m and 15 m?

LENGTH PROBLEMS

(m and cm)

Now try these

Set 1

1 How much steel tape is needed to go round the four sides of a box which is 70 cm long and 48 cm wide?

2 A man is 2 m tall and is 4 cm taller than his son. How tall is the son?

3 It takes 25 cm of string to tie up one bunch of daffodils. How many metres are needed to tie up 72 bunches?

4 A garden is 16 m 65 cm long and 7 m 25 cm wide. Find the length of wire fencing required to go right round the garden.

5 A man buys paving stones 30 cm square to make a path. How many does he need to make a path 15 m long?

6 A man is 49 cm taller than his son, who is 11 cm shorter than his mother. If the man is 1 m 73 cm tall, how tall is the wife?

Set 2

1 A Christmas tree is 1 m 75 cm tall and is planted in a tub which raises it a further 20 cm. If the tub stands on a stool which is 32 cm high, how high is the tree top from the floor?

2 Two boys enter for a 400 m race. One drops out after he has run 230 m while the other boy is 19 m behind him. How much further has the second boy to run?

3 My garden is $14\frac{1}{2}$ m long. Its width is $3\frac{1}{2}$ m less than its length. How far is it all round the garden?

4 A roll of 120 m of cotton material is cut into 6 equal lengths. Each of these is cut to make 5 equal dress lengths. How long is each dress length?

5 One liner is $217\frac{1}{2}$ m long and another is 153 m long. How much longer is the first liner than the second?

6 How much rope is there in 2 lengths each containing 20 m 75 cm?

LENGTH PROBLEMS

(*km and m*)

First try these

Set 1

1 From Jack's house to Jill's house is 893 m. If Jill walks to Jack's house and home again how far short of 2 km does she walk?

2 John rides his bicycle to school, a distance of 3 km. If he returns home to lunch at midday, how far does he ride in going to and from school in a week of 5 days?

3 When Bill left the garage the speedometer of his car read 8516 km. As he entered the garage next it read 9003 km. How far had the car travelled?

4 From Oldtown to Newtown is 85 km measured in a straight line. On a map $\frac{1}{2}$ cm on the map stood for 1 km on the ground. What was the distance on the map between the two towns?

Set 2

1 On country roads our car travels 12 km on 1 litre of petrol. How much petrol will it need on a country journey of 66 km?

2 To one half of 40 km 260 m add twice 10 km 65 m.

3 If Jack walked 8 km, went 40 km by bus and cycled 5500 m, how far did he travel altogether?

4 *A*, *B* and *C* are three ships anchored in a straight line with *C* on the opposite side of *B* to *A*. Ships *B* and *C* are 1000 metres apart. If the distance *AB* is $2\frac{1}{2}$ times the distance *B* to *C*, how many km is it from *A* to *C*?

LENGTH PROBLEMS

(km and m)

Now try these

Set 1

1 When on holiday a family travelled 192 km on the first day, 209 km on the second day, 115 km on the third day and 271 km on the fourth day. How far did they travel altogether?

2 A bus does the journey from town X to town Y, a distance of $87\frac{1}{2}$ km. How far did it travel in (*a*) making a return journey, and (*b*) making the same return journey every day for a week (7 days)?

3 A ship steamed 800 km in 24 hours. If it kept at this speed how far would it travel between midnight on Sunday and midnight the following Tuesday if it travelled in a straight line?

4 Harry and Tom ran in a race. When Harry had run 2 km 840 m he was 360 m ahead of Tom. The latter had 520 m to go to the finishing post? How far had Harry to run?

Set 2

1 The winds, on a certain day, were such that an aeroplane covered the ground at a steady speed of 800 km per hour. How long did it need to travel 5200 km?

2 Find one half of one half of 50 km. From your answer subtract one half of 5000 m. Give your answers in km.

3 X, Y and Z are three houses in a long straight road. From X to Y is 160 m. From Y to Z is three times as far, with Z being on the opposite side of Y to X. How far is it from X to Z and back again?

PRACTICAL EXERCISES IN MEASUREMENT

All can try these

1 Measure the length of your exercise book.

2 Measure the width of your exercise book.

3 What is the distance all round your exercise book?

4 What is the difference between the length and breadth of your exercise book?

5 Measure the length of this book.

6 By how much is the length of this book greater than its breadth?

7 Find the distance all round this book.

8 Draw a line on your paper and measure its length.

9 Measure the length of your desk.

10 How long is your pen (or pencil)?

11 Draw a rectangle 9 cm long and 6 cm wide. What is the distance all round it?

A blackboard is $6\frac{1}{2}$ m long and $2\frac{1}{2}$ m wide. Draw a rectangle $6\frac{1}{2}$ cm by $2\frac{1}{2}$ cm to represent the board.

12 What is the distance round the rectangle which you have drawn, and how many metres does this represent?

13 Choose a pane of glass in your classroom window and measure its length and breadth.

14 What is the width of your classroom door?

15 Measure the length and breadth of your classroom. What is the distance all round the room?

16 What is the distance across your school hall?

17 Measure the width of the door of your classroom cupboard.

18 How high is your classroom door?

19 Measure the distance from one corner of a book to the opposite corner. The line joining the opposite corners of a square or rectangle is called a **diagonal.**

SHAPES AND THEIR COMPOSITION

All can try these

1 Draw a square of side 8 cm, and copy the pattern shown.

If the square were cut in half, the pattern of each of the halves would be exactly alike.

Patterns like this, which are exactly balanced, are called **symmetrical.**

2 We know that a square has four sides.
Look at these drawings of triangles. A triangle has three sides.

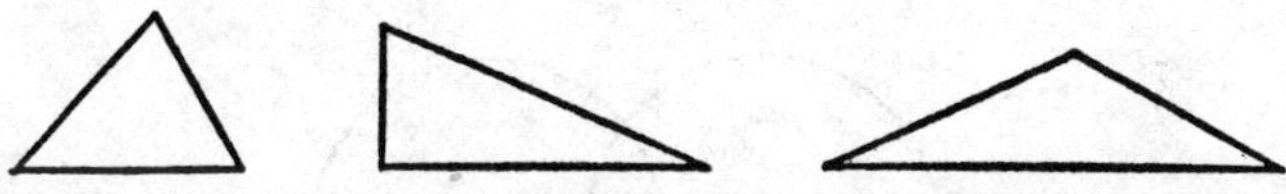

How many squares and how many triangles can you find in the square at the top of the page?

3 Which shapes, **a—m**, are parts of the church, the ship, the dog?

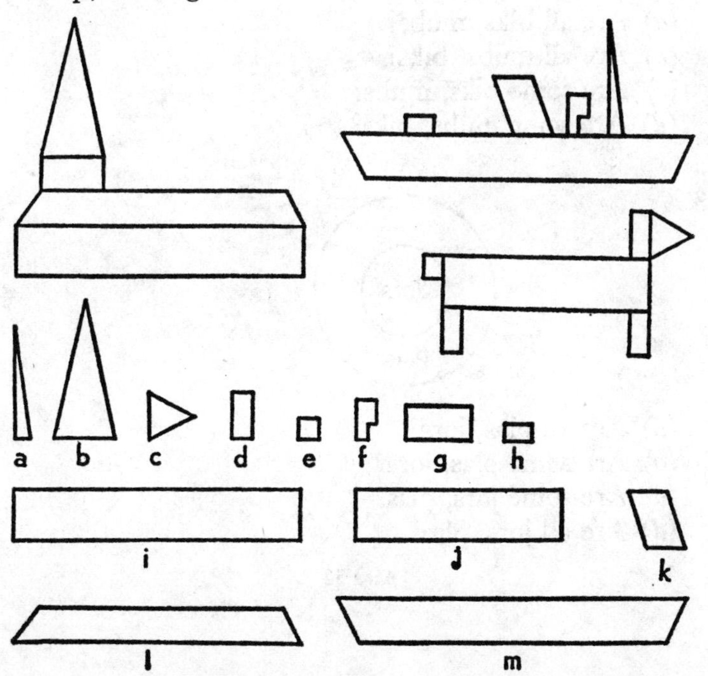

WORK WITH SETS

First try these

1

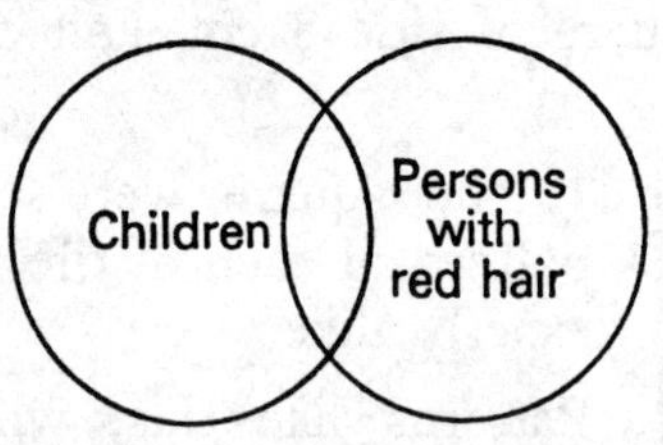

(*a*) Are some children, persons with red hair?
(*b*) Are all persons with red hair, children?
(*c*) Are some persons with red hair, children?
(*d*) Are all children, persons with red hair?

2

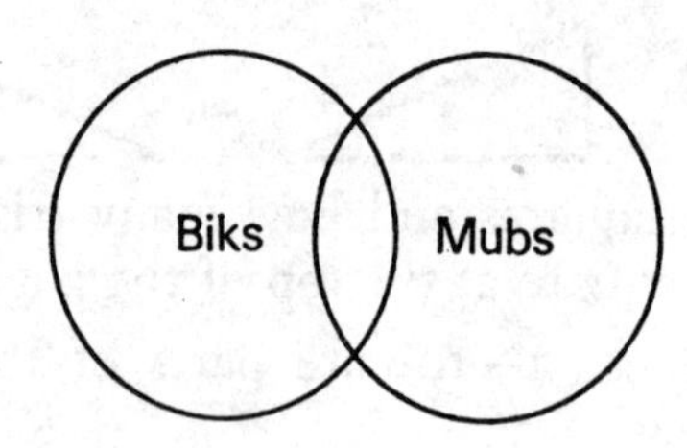

(*a*) Are all biks, mubs?
(*b*) Are all mubs, biks?
(*c*) Are some biks, mubs?
(*d*) Are some mubs, biks?

3

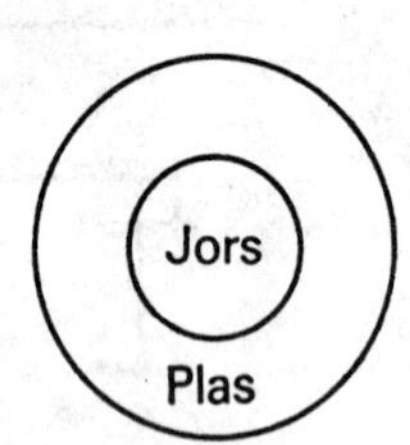

(*a*) Are all plas, jors?
(*b*) Are some plas, jors?
(*c*) Are some jors, plas?
(*d*) Are all jors, plas?

WORK WITH SETS

Now try these

1

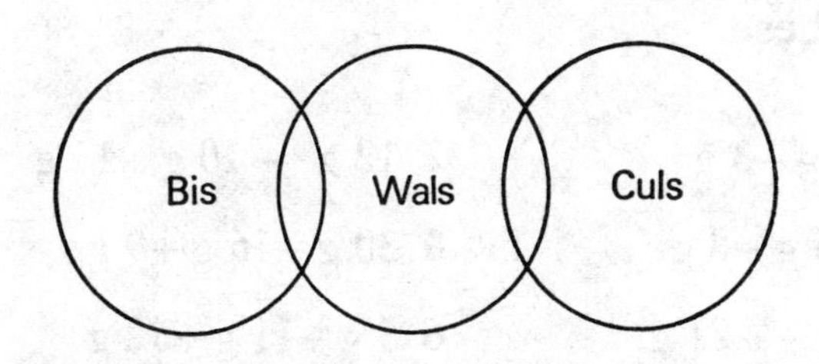

(*a*) Are some wals, culs?
(*b*) Are all bis, wals?
(*c*) Are some culs, bis?
(*d*) Are all culs, wals?
(*e*) Are some bis, culs?
(*f*) Are all culs, bis?
(*g*) Are some culs, wals?

2

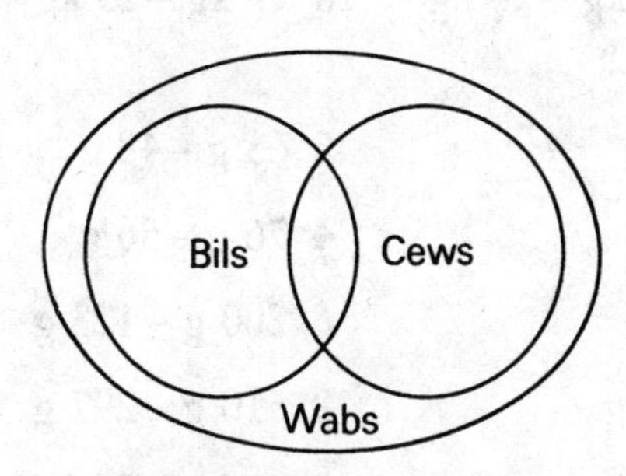

(*a*) Are all bils, wabs?
(*b*) Are some cews, bils?
(*c*) Are all wabs, cews?
(*d*) Are all cews, wabs?
(*e*) Are some wabs, bils?
(*f*) Are some bils, cews?

WEIGHT ADDITION AND SUBTRACTION

First try these

Set 1

1 7 g + 4 g + 3 g

2 12 g + 10 g + 15 g

3 2 g + 25 g + 8 g

4 30 g + 16 g + 9 g

5 17 g + 1 g + 21 g

6 5 g + 11 g + 12 g

7 3 kg + 1 kg + 7 kg

8 4 kg + 8 kg + 13 kg

9 5 kg + 9 kg + 14 kg

10 20 kg + 7 kg + 11 kg

11 $4\frac{1}{2}$ kg + 8 kg + $1\frac{1}{2}$ kg

12 16 kg + 5 kg + 2 kg

13 19 kg + 2 kg + 26 kg

14 $23\frac{1}{2}$ kg + 10 kg + $6\frac{1}{2}$ kg

15 $17\frac{1}{2}$ kg + $5\frac{1}{2}$ kg + 9 kg

16 14 kg + 22 kg + $18\frac{1}{2}$ kg

17 30 kg + $27\frac{1}{2}$ kg + $2\frac{1}{2}$ kg

18 $7\frac{1}{2}$ kg + 23 kg + $11\frac{1}{2}$ kg

Set 2

1 37 g - 25 g

2 63 g - 48 g

3 49 g - 30 g

4 76 g - 56 g

5 100 g - 52 g

6 200 g - 123 g

7 426 g - 359 g

8 510 g - 297 g

9 $28\frac{1}{2}$ kg - 13 kg

10 47 kg - 26 kg

11 63 kg - $45\frac{1}{2}$ kg

12 $34\frac{1}{2}$ kg - $18\frac{1}{2}$ kg

13 52 kg - $19\frac{1}{2}$ kg

14 60 kg - 7 kg

15 45 kg - $24\frac{1}{2}$ kg

16 100 kg - 58 kg

17 $92\frac{1}{2}$ kg - 68 kg

18 85 kg - $49\frac{1}{2}$ kg

WEIGHT ADDITION AND SUBTRACTION

Now try these

Set 1

1 16 g + 23 g + 12 g

2 19 g + 36 g + 7 g

3 34 g + 10 g + 21 g

4 42 g + 13 g + 28 g

5 15 g + 71 g + 26 g

6 105 g + 64 g + 83 g

7 8 kg + 20 kg + 11 kg

8 10 kg + 14 kg + 10 kg

9 25 kg + 3 kg + 14 kg

10 32 kg + $40\frac{1}{2}$ kg + 6 kg

11 $18\frac{1}{2}$ kg + 15 kg + $36\frac{1}{2}$ kg

12 $56\frac{1}{2}$ kg + $31\frac{1}{2}$ kg + $28\frac{1}{2}$ kg

13 $47\frac{1}{2}$ kg + $19\frac{1}{2}$ kg + $55\frac{1}{2}$ kg

14 29 kg + $59\frac{1}{2}$ kg + 50 kg

15 $54\frac{1}{2}$ kg + $37\frac{1}{2}$ kg + 12 kg

16 $85\frac{1}{2}$ kg + 94 kg + $41\frac{1}{2}$ kg

17 30 kg + $63\frac{1}{2}$ kg + $49\frac{1}{2}$ kg

18 $46\frac{1}{2}$ kg + 83 kg + $105\frac{1}{2}$ kg

Set 2

1 84 g − 37 g

2 150 g − 108 g

3 300 g − 85 g

4 529 g − 456 g

5 706 g − 283 g

6 614 g − 340 g

7 831 g − 697 g

8 999 g − 638 g

9 40 kg − $9\frac{1}{2}$ kg

10 $35\frac{1}{2}$ kg − 17 kg

11 $82\frac{1}{2}$ kg − $36\frac{1}{2}$ kg

12 93 kg − $44\frac{1}{2}$ kg

13 104 kg − 15 kg

14 $157\frac{1}{2}$ kg − 95 kg

15 96 kg − $32\frac{1}{2}$ kg

16 134 kg − 57 kg

17 257 kg − 168 kg

18 308 kg − $198\frac{1}{2}$ kg

WEIGHT ADDITION AND SUBTRACTION

First try these

Set 1

1

	kg	g
	4	300
+	2	50

2

	kg	g
	1	270
+	6	120

3

	kg	g
	3	95
+	7	430

4

	kg	g
	8	405
+	5	127

5

	kg	g
	11	623
+		59

6

	kg	g
	4	365
+	8	528

7

	kg	g
	6	290
+	3	720

8

	kg	g
	10	680
+	12	455

9

	kg	g
	14	390
+	1	986

Set 2

1

	kg	g
	5	70
−	2	40

2

	kg	g
	3	290
−	1	85

3

	kg	g
	9	680
−	6	204

4

	kg	g
	17	360
−	7	109

5

	kg	g
	13	820
−	6	470

6

	kg	g
	11	532
−	3	85

7

	kg	g
	20	421
−	11	283

8

	kg	g
	16	945
−	10	761

9

	kg	g
	15	130
−	4	810

10

	kg	g
	17	250
−	9	750

11

	kg	g
	18	827
−	14	603

12

	kg	g
	23	440
−	19	782

WEIGHT ADDITION AND SUBTRACTION

Now try these

Set 1

1

kg	g
12	219
+ 5	37

2

kg	g
18	76
+ 11	431

3

kg	g
29	512
+ 14	392

4

kg	g
31	676
+	324

5

kg	g
51	817
+ 35	249

6

kg	g
63	764
+ 25	682

7

kg	g
58	939
+ 47	514

8

kg	g
84	620
+ 117	409

9

kg	g
97	32
+ 121	986

Set 2

1

kg	g
7	432
– 1	296

2

kg	g
12	205
– 9	513

3

kg	g
30	365
– 18	728

4

kg	g
24	38
– 19	195

5

kg	g
49	637
– 43	983

6

kg	g
87	610
– 54	863

7

kg	g
60	400
– 38	791

8

kg	g
95	750
– 89	162

9

kg	g
110	618
– 94	809

10

kg	g
198	106
– 188	196

11

kg	g
275	23
– 190	517

12

kg	g
326	826
– 153	827

WEIGHT MULTIPLICATION

First try these

Set 1

1 g
8
× 2

2 g
17
× 2

3 g
41
× 2

4 g
59
× 2

5 kg
3
× 2

6 kg
7
× 2

7 kg
$8\frac{1}{2}$
× 2

8 kg
10
× 2

9 kg
$6\frac{1}{2}$
× 2

10 kg
12
× 2

11 kg
$15\frac{1}{2}$
× 2

12 kg
$27\frac{1}{2}$
× 2

Set 2

1 kg g
2 30
× 2

2 kg g
5 100
× 2

3 kg g
10 230
× 2

4 kg g
7 450
× 2

5 kg g
12 85
× 2

6 kg g
9 362
× 2

7 kg g
11 190
× 2

8 kg g
6 417
× 2

9 kg g
14 193
× 2

10 kg g
8 550
× 2

11 kg g
15 600
× 2

12 kg g
13 476
× 2

WEIGHT MULTIPLICATION

Now try these

Set 1

1	2	3	4
g	g	g	g
65	93	187	462
$\times 2$	$\times 2$	$\times 2$	$\times 2$

5	6	7	8
kg	kg	kg	kg
$7\frac{1}{2}$	19	$48\frac{1}{2}$	$65\frac{1}{2}$
$\times 2$	$\times 2$	$\times 2$	$\times 2$

9	10	11	12
kg	kg	kg	kg
$80\frac{1}{2}$	135	264	$401\frac{1}{2}$
$\times 2$	$\times 2$	$\times 2$	$\times 2$

Set 2

1 kg	g	2 kg	g	3 kg	g
9	280	11	375	8	492
	$\times 2$		$\times 2$		$\times 2$

4 kg	g	5 kg	g	6 kg	g
17	500	29	407	20	681
	$\times 2$		$\times 2$		$\times 2$

7 kg	g	8 kg	g	9 kg	g
32	715	43	264	58	870
	$\times 2$		$\times 2$		$\times 2$

10 kg	g	11 kg	g	12 kg	g
76	926	95	568	103	698
	$\times 2$		$\times 2$		$\times 2$

WEIGHT DIVISION

First try these

Set 1

Divide the following quantities by 2

1 12 g

2 40 g

3 64 g

4 28 g

5 76 g

6 100 g

7 182 g

8 150 g

9 14 kg

10 18 kg

11 36 kg

12 42 kg

13 23 kg

14 19 kg

15 39 kg

16 67 kg

Set 2

Divide the following quantities by 2

1 4 kg 20 g

2 6 kg 80 g

3 2 kg 140 g

4 10 kg 300 g

5 8 kg 460 g

6 16 kg 224 g

7 20 kg 780 g

8 18 kg 670 g

9 12 kg 946 g

10 14 kg 510 g

11 3 kg 10 g

12 5 kg 60 g

13 1 kg 240 g

14 9 kg 400 g

15 7 kg 816 g

16 22 kg 350 g

WEIGHT DIVISION

Now try these

Set 1

Divide the following quantities by 2

1	98 g	**2**	162 g
3	392 g	**4**	506 g
5	714 g	**6**	932 g
7	49 kg	**8**	102 g
9	456 kg	**10**	395 kg
11	1000 kg	**12**	823 kg
13	2010 kg	**14**	1596 kg
15	4172 kg	**16**	3708 kg

Set 2

Divide the following quantities by 2

1	6 kg 150 g	**2**	11 kg 64 g
3	17 kg 200 g	**4**	15 kg 392 g
5	23 kg 404 g	**6**	30 kg 906 g
7	48 kg 560 g	**8**	41 kg 728 g
9	39 kg 98 g	**10**	57 kg 682 g
11	46 kg 824 g	**12**	69 kg 576 g
13	77 kg 978 g	**14**	85 kg 10 g
15	93 kg 716 g	**16**	91 kg 832 g

WEIGHT PROBLEMS

First try these

Set 1

1 A boy weighs 40 kg 250 g and his sister 25 kg 700 g. How much heavier is the boy than his sister?
2 A woman bought $1\frac{1}{2}$ kg potatoes, $\frac{1}{2}$ kg carrots, 1 kg tomatoes and bananas weighing 750 g. Find the total weight.
3 A farmer shares 34 kg apples among 8 boys. What weight does each get?
4 To make a Christmas pudding 40 g fruit are needed. How much fruit is needed for 4 puddings of the same size?
5 A basket of fruit weighs 3 kg 60 g. If the weight of the empty basket is 850 g, what is the weight of the fruit?
6 How many 6 kg lots of potatoes can be weighed up from 48 kg?
7 If 960 kg of cement is made up into 12 kg bags, how many will be made?

Set 2

1 A package weighs 76 kg and a second package $41\frac{1}{2}$ kg. How much heavier is the first than the second?
2 A cheese weighs 15 kg. A piece weighing $5\frac{1}{2}$ kg is cut from it. What is the weight of cheese left?
3 Find the cost of 1 kg at 1p a gramme.
4 How many $\frac{1}{2}$ kg packets of butter can be made up from 24 kg?
5 What is the cost of a jar of sweets weighing 3 kg, if the sweets cost 20p per $\frac{1}{4}$ kg?
6 How many bags weighing 50 kg can be made up from 1000 kg?
7 Multiply $17\frac{1}{2}$ kg by 2.
8 A cake weighs $1\frac{1}{2}$ kg, and 250 g of icing is added to it. Find the total weight of the cake.

WEIGHT PROBLEMS

Now try these

Set 1

1 Three boxes of plums weighed 3 kg 450 g each. If the weight of the three empty boxes together was 925 g, what was the weight of the plums?
2 Five hundred kilogrammes of coal are to be made up into 40 kg sacks. How many whole sacks will it make?
3 At the grocer's Mother bought 3 kg flour, 500 g butter, 5 kg sugar and 850 g biscuits. She also bought 600 g bacon. What was the total weight of goods bought?
4 What do 14 kg sweets cost at 45p a half-kilogramme?
5 A nurseryman buys 1200 kg of seed potatoes and makes them up into 24 kg bags. How many bags does he fill?
6 One bale of wool weighs 70 kg and another weighs $64\frac{1}{2}$ kg. What was their total weight?
7 A greengrocer buys five 50 kg sacks of potatoes. During one morning he sells 95 kg. How many kg are left?

Set 2

1 A confectioner buys 1 dozen boxes of chocolates which weigh 500 g each. Find the total weight in kg.
2 If 28 kg fruit is divided equally among 8 boys, how much will each receive?
3 Find the cost of 10 kg at £0·35 per kg.
4 Multiply 3 kg 500 g by 2.
5 A wedding cake weighs $4\frac{1}{2}$ kg. The icing and other decorations weigh 1 kg 750 g. What is the weight of the cake itself?
6 A turkey weighed $7\frac{1}{2}$ kg. When prepared for the oven it weighed 5 kg 250 g. Find the weight of the waste material.
7 A dairyman has $9\frac{1}{2}$ kg of butter. He sells 4 slabs each weighing $\frac{1}{2}$ kg. How much has he left?

GENERAL REVISION

First try these

Add

1 £

$$\begin{array}{r} 4{\cdot}16 \\ 0{\cdot}07\tfrac{1}{2} \\ +16{\cdot}29 \\ \hline \end{array}$$

2 km m

$$\begin{array}{rr} 14 & 295 \\ 9 & 784 \\ + & 867 \\ \hline \end{array}$$

3

$$\begin{array}{r} 1968 \\ 2047 \\ +\ 532 \\ \hline \end{array}$$

Subtract

4

$$\begin{array}{r} 4075 \\ -2986 \\ \hline \end{array}$$

5 £

$$\begin{array}{r} 5{\cdot}00 \\ -2{\cdot}93 \\ \hline \end{array}$$

6 m cm

$$\begin{array}{rr} 9 & 25 \\ -1 & 75 \\ \hline \end{array}$$

Multiply

7 km m

$$\begin{array}{rr} 3 & 400 \\ & \times 2 \\ \hline \end{array}$$

8 kg g

$$\begin{array}{rr} 11 & 120 \\ & \times 2 \\ \hline \end{array}$$

9 £

$$\begin{array}{r} 2{\cdot}81 \\ \times 6 \\ \hline \end{array}$$

Divide

10 16 kg 826 g ÷ 2 **11** £0·45 ÷ 5 **12** 49 kg ÷ 7

13 Write in figures three thousand one hundred and six.

14 Add together 583, 1044 and 29, and subtract the result from 5000.

15 What is the cost of 60 g at 2p per g?

16 What must be added to 289 cm to make 4 m?

17 How many oranges will there be in 38 boxes if each contains 3 dozen?

18 How many 200 g packets of sweets can be made up from 5 kg?

19 What is the difference between 450 cm and 5 m?

20 A man has to fill 18 boxes with 96 eggs in each. How many eggs does he need?

GENERAL REVISION

Now try these

Add

1

m	cm
4	22
6	25
+	17

2

kg	g
15	500
6	750
+2	250

3

£
$103 \cdot 12\frac{1}{2}$
$0 \cdot 86$
$+\ 84 \cdot 59\frac{1}{2}$

Subtract

4

£
$129 \cdot 01\frac{1}{2}$
$-\ 59 \cdot 83$

5

kg	g
30	250
−7	400

6

km	m
84	405
−18	860

Multiply

7

$$\begin{array}{r} 95 \\ \times 86 \end{array}$$

8

m	cm
63	54
	×2

9

$$\begin{array}{r} \text{kg} \\ 98\frac{1}{2} \\ \times 2 \end{array}$$

Divide

10 £254·70 ÷ 10 **11** 948 ÷ 45 **12** 61 kg 380 g ÷ 2

13 Write in figures four thousand seven hundred and three.

14 In one bag there are 6 kg of potatoes and a second bag weighs 2 kg more. Find the total weight of both bags.

15 A woman earns 70p per hour. How much will she earn (*a*) in a day of 8 hours, and (*b*) in a week of 24 hours?

16 Multiply 3 m 65 cm by 2, and add 29 cm to the result.

17 Find one half of the sum of 17, 215, 124, 63, 59, 84.

18 A ship is 250 m long, but the side of the dock where it is tied up is only $236\frac{1}{2}$ m long. How much longer is the ship than the dockside?

19 A wheel is 43 cm round. How many times will it turn in going 1333 cm?

20 From a certain number 36 was subtracted, and the result was divided by 4. The answer was 16. What was the number?

CAPACITY ADDITION

First try these

Set 1

1 2 litres + 3 litres

2 6 litres + 1 litre

3 4 litres + 5 litres + 1 litre

4 7 litres + 4 litres + 2 litres

5 $3\frac{1}{2}$ litres + $2\frac{1}{2}$ litres + 6 litres

6 9 litres + 5 litres + $4\frac{1}{2}$ litres

7 8 litres + $7\frac{1}{2}$ litres + 1 litre

8 12 litres + $7\frac{1}{2}$ litres + 2 litres

Set 2

Give the answer in litres and ml when necessary

1 80 ml + 60 ml

2 20 ml + 150 ml

3 261 ml + 100 ml

4 180 ml + 320 ml

5 500 ml + 500 ml

6 250 ml + 750 ml

7 800 ml + 210 ml

8 700 ml + 490 ml

Set 3

1

litres	ml
4	80
+ 5	90

2

litres	ml
3	110
+ 2	160

3

litres	ml
7	193
+ 1	305

4

litres	ml
6	429
+ 8	514

5

litres	ml
8	500
+ 10	620

6

litres	ml
9	840
+ 11	370

CAPACITY ADDITION

Now try these

Set 1

1 4 litres + 15$\frac{1}{2}$ litres + 13 litres

2 12 litres + 3$\frac{1}{2}$ litres + 10$\frac{1}{2}$ litres

3 19 litres + 20$\frac{1}{2}$ litres + 11 litres

4 14$\frac{1}{2}$ litres + 17 litres + 29 litres

5 24 litres + 5$\frac{1}{2}$ litres + 16$\frac{1}{2}$ litres

6 38 litres + 6$\frac{1}{2}$ litres + 25 litres

7 31 litres + 40 litres + 23$\frac{1}{2}$ litres

8 48$\frac{1}{2}$ litres + 30$\frac{1}{2}$ litres + 57$\frac{1}{2}$ litres

Set 2

Give your answer in litres and ml when necessary

1 600 ml + 392 ml

2 450 ml + 580 ml

3 861 ml + 295 ml

4 906 ml + 279 ml

5 763 ml + 237 ml

6 809 ml + 652 ml

7 954 ml + 928 ml

8 560 ml + 439 ml

Set 3

1

litres	ml
14	375
+ 9	625

2

litres	ml
23	489
+ 17	254

3

litres	ml
46	802
+ 39	530

4

litres	ml
97	968
+ 55	389

5

litres	ml
107	421
+ 96	837

6

litres	ml
175	893
+ 64	415

CAPACITY SUBTRACTION

First try these

Set 1

1 12 litres – 7 litres

2 13 litres – $1\frac{1}{2}$ litres

3 9 litres – $2\frac{1}{2}$ litres

4 $10\frac{1}{2}$ litres – $3\frac{1}{2}$ litres

5 $17\frac{1}{2}$ litres – 5 litres

6 22 litres – 12 litres

7 14 litres – $6\frac{1}{2}$ litres

8 $16\frac{1}{2}$ litres – $9\frac{1}{2}$ litres

9 $20\frac{1}{2}$ litres – $11\frac{1}{2}$ litres

10 18 litres – $13\frac{1}{2}$ litres

Set 2

1 45 ml – 20 ml

2 200 ml – 120 ml

3 350 ml – 230 ml

4 500 ml – 250 ml

5 750 ml – 500 ml

6 640 ml – 580 ml

7 800 ml – 60 ml

8 700 ml – 340 ml

9 950 ml – 500 ml

10 830 ml – 270 ml

Set 3

1

	litres	ml
	2	183
–	1	150

2

	litres	ml
	4	95
–	3	30

3

	litres	ml
	6	370
–	4	190

4

	litres	ml
	7	420
–	3	70

5

	litres	ml
	8	630
–	2	580

6

	litres	ml
	10	900
–	7	830

CAPACITY SUBTRACTION

Now try these

Set 1

1 29 litres – 18½ litres

2 58 litres – 40 litres

3 73½ litres – 47 litres

4 49½ litres – 16 litres

5 68½ litres – 35½ litres

6 75½ litres – 31 litres

7 54 litres – 26 litres

8 93 litres – 69½ litres

9 107 litres – 85½ litres

10 151½ litres – 101½ litres

Set 2

1 610 ml – 590 ml

2 372 ml – 184 ml

3 307 ml–198 ml

4 425 ml – 179 ml

5 867 ml – 549 ml

6 682 ml – 291 ml

7 938 ml – 754 ml

8 728 ml – 456 ml

9 897 ml – 679 ml

10 908 ml – 861 ml

Set 3

1

	litres	ml
	11	230
–	4	162

2

	litres	ml
	15	174
–	13	297

3

	litres	ml
	49	383
–	17	975

4

	litres	ml
	54	296
–	48	287

5

	litres	ml
	107	100
–	68	394

6

	litres	ml
	90	805
–	73	964

CAPACITY MULTIPLICATION AND DIVISION

First try these

Set 1

Multiply the following quantities by 2

1 9 litres
2 7 litres
3 $11\frac{1}{2}$ litres
4 15 litres
5 $19\frac{1}{2}$ litres
6 24 litres
7 27 litres
8 $31\frac{1}{2}$ litres
9 3 litres 20 ml
10 4 litres 130 ml
11 2 litres 90 ml
12 6 litres 325 ml
13 5 litres 186 ml
14 13 litres 500 ml
15 9 litres 270 ml
16 7 litres 480 ml
17 11 litres 461 ml
18 19 litres 625 ml

Set 2

Divide the following quantities by 2

1 21 litres
2 27 litres
3 48 litres
4 60 litres
5 35 litres
6 53 litres
7 86 litres
8 71 litres
9 8 litres 400 ml
10 6 litres 350 ml
11 10 litres 860 ml
12 5 litres 100 ml
13 3 litres 70 ml
14 9 litres 200 ml
15 11 litres 190 ml
16 15 litres 280 ml
17 14 litres 532 ml
18 17 litres 306 ml

CAPACITY MULTIPLICATION AND DIVISION

Now try these

Set 1

Multiply the following quantities by 2

1 34 litres
2 52 litres
3 62 litres
4 77 litres
5 125 litres
6 348 litres
7 267 litres
8 523 litres
9 16 litres 250 ml
10 19 litres 380 ml
11 11 litres 621 ml
12 25 litres 417 ml
13 43 litres 763 ml
14 76 litres 682 ml
15 108 litres 975 ml
16 124 litres 809 ml
17 99 litres 842 ml
18 85 litres 721 ml

Set 2

Divide the following quantities by 2

1 38 litres
2 68 litres
3 87 litres
4 105 litres
5 134 litres
6 397 litres
7 483 litres
8 902 litres
9 18 litres 56 ml
10 39 litres 110 ml
11 135 litres 172 ml
12 94 litres 238 ml
13 297 litres 406 ml
14 123 litres 670 ml
15 326 litres 824 ml
16 505 litres 938 ml
17 519 litres 716 ml
18 611 litres 872 ml

CAPACITY PROBLEMS

First try these

Set 1

1. A milk churn holds 55 litres. From a churn a milkman sold $30\frac{1}{2}$ litres. How much milk had he left?
2. Multiply 2 litres 500 ml by one half of four.
3. From the sum of $8\frac{1}{2}$ litres and $3\frac{1}{2}$ litres, take $7\frac{1}{2}$ litres.
4. A baker, using the same amount of milk each day, used 278 litres in 2 days. How much did he use per day?
5. In three days a grocer sold the following amounts of vinegar: $15\frac{1}{2}$ litres, 21 litres, $25\frac{1}{2}$ litres. How much was sold altogether?
6. A herd of cows gives 8 churns of milk, each holding 45 litres. If this milk is shared equally between two dairies, how much does each take?
7. At a garage there was a tank holding 1000 litres of oil. When 721 litres 500 ml of this had been sold, how much was left?
8. How many litres of liquid are needed to fill 80 bottles each holding 50 ml?

Set 2

1. A barrel of tar contains 128 litres. How much is left after using 6 bucketfuls, if each bucket holds 10 litres?
2. A litre of milk will fill two bottles. How many bottles of this size are needed for 420 litres?
3. If 21 litres of oil will fill the tank of an oil stove 12 times, how much oil does the tank hold?
4. A motor lorry carried a tank containing 2000 litres of paraffin. How much remained in the tank after 5 shops had each been supplied with 300 litres, and allowing for 2 litres waste?
5. If a litre of water weighs 1 kg, what is the weight of (*a*) $\frac{1}{2}$ litre of water, and (*b*) 6 litres of water?
6. How much ice-cream must be ordered for a dance attended by 200 people, if 1 litre of ice-cream is needed for every 8 people?

CAPACITY PROBLEMS

Now try these

Set 1

1. A shopkeeper sells to three customers 15 litres, $1\frac{1}{2}$ litres and 4 litres. Find the total cost at 2p per litre.
2. A well bucket holds $17\frac{1}{2}$ litres of water. How much water is drawn off if the bucket is filled 11 times?
3. A milkman sold 35 litres 400 ml of milk from a 50 litre churn. How much milk had he left?
4. How many half-litre mugs can I fill from $6\frac{1}{2}$ litres?
5. Every day 25 litres of liquid soap is used in a factory. How much is used in (*a*) 5 days, (*b*) 30 days?
6. Ten cans of water, each holding $15\frac{1}{2}$ litres, are poured into a barrel already holding 35 litres. How much is now in the barrel?
7. A garage has three tanks containing 900, 1250 and 975 litres of paraffin. If 1075 litres are sold, how much is left?
8. A farmer's cows gave 675 litres of milk. How many 60 litre churns would be needed to hold this milk, and how much would be left over?

Set 2

1. In 5 days a shop sold 104, 93, 116, 87 and 95 litres of milk. If it was sold in $\frac{1}{2}$-litre bottles, how many bottles were needed?
2. Subtract $15\frac{1}{2}$ litres from one half the sum of 49 litres and 26 litres.
3. An oil merchant buys 5 drums of oil each holding 60 litres. On a winter's day he sells 100 litres and 20 half-litres of oil. How much has he left?
4. A farmer's herd of cows produced 160 litres of milk a day. If he used 30 litres in feeding his family and some young calves, how much did he sell each day?
5. A dairyman delivers 7 crates of milk daily to a hotel. If each crate contains thirty $\frac{1}{2}$-litre bottles, how many litres of milk does he deliver daily?

GENERAL REVISION

First try these

Add

1

4923
158
+27

2

m	cm
2	53
4	10
+3	95

3

litres	ml
35	600
20	300
+ 6	900

Subtract

4

kg	g
104	800
−76	950

5

8010
−5293

6

km	m
80	390
−46	700

Multiply

7 76 × 34

8 5 litres × 8

9 £9·35 × 12

Divide

10 8 litres 200 ml ÷ 4

11 735 ÷ 35

12 13 m 40 cm ÷ 2

13 Write in figures five thousand and twenty-nine.

14 If I pay £3·00 for 5 m of cloth, how much shall I pay for 20 m?

15 A church steeple is 50 m 75 cm high. A steeplejack climbs to within $7\frac{1}{2}$ m of the top. How high is he above the ground?

16 A greengrocer has 16 kg of apples. He sells half of them at 20p per kg and the remainder at 16p per kg. What were his total takings?

17 We have twelve dozen tulips to divide equally among 16 vases. How many tulips should we put in each vase?

18 A boy lost half his marbles in a game. In a second game he lost half of those he had at the end of the first game. He finished with 4. How many marbles had he to begin with?

19 A box holds ten 600 ml bottles of a liquid. If 60 boxes are loaded on to a van, how many litres of liquid are loaded?

GENERAL REVISION

Now try these

Add

1

£
49·91
25·68
+ 90·75

2

kg	g
290	450
3	730
+ 48	560

3

litres	ml
73	350
9	860
+	490

Subtract

4

km	m
103	0
− 58	400

5

m	cm
75	68
− 29	84

6

9013
− 2905

Multiply

7 $9\frac{1}{2}$ kg × 10

8 91 × 89

9 4 litres 600 ml × 8

Divide

10 108 litres ÷ 9

11 1637 ÷ 52

12 £58·68 ÷ 12

13 Write in figures eight thousand and eight.

14 A shopkeeper sold 600 ml of paraffin to each of 5 persons. If he started the day with 10 litres how much had he left?

15 A bucket holds $20\frac{1}{2}$ kg of sand. What weight of sand can be put into 4 buckets?

16 Divide £1·00 between 2 girls so that one has 20p more than the other.

17 A boy drinks 500 ml of milk every day. How many litres will he drink in (*a*) 7 days, and (*b*) a month of 28 days?

18 How many times can 59 be subtracted from 1250? What is the remainder?

19 A man aged 72 died in 1971. The man's father was 35 when the man was born. In what year was the father born?

20 What change should I have from £100·00 after buying 2 articles at £37·85 each?

CHANGING UNITS

First try these

Change these **minutes** into **hours**

1 76 **2** 124 **3** 98 **4** 198 **5** 67 **6** 143

Change these **pence** into **pounds**

7 90 **8** 130 **9** 147 **10** 180 **11** 179 **12** 200

Change these **hours** into **days**

13 65 **14** 74 **15** 91 **16** 120 **17** 145 **18** 180

Change these **days** into **weeks**

19 18 **20** 36 **21** 51 **22** 70 **23** 85 **24** 100

Change these **metres** into **centimetres**

25 20 **26** 40 **27** 54 **28** 70 **29** 86 **30** 101

Change these **litres** into **millilitres**

31 10 **32** 19 **33** 27 **34** 36 **35** 62 **36** 74

Change these **half-litres** into **litres**

37 8 **38** 12 **39** 16 **40** 7 **41** 28 **42** 19

Change these **kilogrammes** into **grammes**

43 6 **44** 12 **45** 18 **46** 29 **47** 34 **48** 48

Change these **kilometres** into **metres**

49 18 **50** 30 **51** 46 **52** 56 **53** 68 **54** 80

Change these **centimetres** into **metres**

55 134 **56** 132 **57** 162 **58** 371 **59** 485 **60** 501

CHANGING UNITS

Now try these

Change these **minutes** into **hours**

1 168 **2** 136 **3** 208 **4** 245 **5** 96 **6** 184

Change these **pence** into **pounds**

7 94 **8** 189 **9** 141 **10** 220 **11** 260 **12** 300

Change these **hours** into **days**

13 83 **14** 98 **15** 109 **16** 126 **17** 164 **18** 197

Change these **days** into **weeks**

19 16 **20** 49 **21** 74 **22** 87 **23** 99 **24** 126

Change these **kilogrammes** into **grammes**

25 18 **26** 47 **27** 58 **28** 92 **29** 108 **30** 120

Change these **litres** into **millilitres**

31 9 **32** 21 **33** 48 **34** 77 **35** 102 **36** 130

Change these **half-litres** into **litres**

37 17 **38** 29 **39** 45 **40** 36 **41** 63 **42** 78

Change these **grammes** into **kilogrammes**

43 1754 **44** 2805 **45** 4167 **46** 3952 **47** 8318 **48** 9683

Change these **millilitres** into **litres**

49 3875 **50** 4002 **51** 7617 **52** 5205 **53** 6912 **54** 7051

Change these **centimetres** into **metres**

55 151 **56** 269 **57** 281 **58** 4105 **59** 3132 **60** 6154

TIME ADDITION

First try these

No.	Units		First	Add
1	h	min	1 20	+2 20
2	h	min	2 15	+4 20
3	h	min	3 30	+2 10
4	h	min	1 10	+3 15
5	h	min	5 40	+1 25
6	h	min	7 50	+15
7	h	min	2 40	+3 25
8	h	min	4 45	+4 25
9	h	min	3 40	+1 30
10	h	min	5 35	+6 30
11	h	min	1 45	+4 20
12	h	min	7 55	+7 45
13	min	sec	7 40	+2 20
14	min	sec	3 50	+4 15
15	min	sec	8 45	+1 25
16	min	sec	4 55	+35
17	days	h	3 10	+2 6
18	days	h	8 12	+2 9
19	days	h	6 10	+4 12
20	days	h	4 18	+4 3
21	days	h	3 12	+1 16
22	days	h	1 18	+5 9
23	days	h	4 21	+2 13
24	days	h	3 16	+9 10
25	days	h	14 23	+5
26	days	h	13 20	+2 7
27	days	h	15 18	+4 15
28	days	h	16 19	+5 12

TIME ADDITION

Now try these

	h	min		h	min		h	min		h	min
1	9	15	**2**	5	10	**3**	12	30	**4**	13	45
	+4	20		+2	35		+11	45		+9	40

	h	min		h	min		h	min		h	min
5	16	35	**6**	23	40	**7**	12	55	**8**	18	50
	+8	40		+4	50		+20	45		+14	25

	h	min		h	min		h	min		h	min
9	19	28	**10**	33	16	**11**	22	51	**12**	31	19
	+7	42		+11	53		+6	27		+9	52

	min	sec		min	sec		min	sec		min	sec
13	15	37	**14**	24	30	**15**	32	15	**16**	52	53
	+21	25		+13	48		+10	45		+1	31

	days	h		days	h		days	h		days	h
17	8	15	**18**	15	21	**19**	32	16	**20**	12	8
	+14	10		+21	12		+18	9		+29	17

	days	h		days	h		days	h		days	h
21	7	0	**22**	26	17	**23**	35	21	**24**	21	23
	+18	14		+4	20		+5	13		+9	22

	days	h		days	h		days	h		days	h
25	39	16	**26**	28	17	**27**	19	19	**28**	37	22
	+4	19		+15	23		+27	21		+8	15

TIME SUBTRACTION

First try these

1	h	min	**2**	h	min	**3**	h	min	**4**	h	min
	3	20		5	45		4	50		3	40
	−1	10		−1	25		−2	10		−2	20

5	h	min	**6**	h	min	**7**	h	min	**8**	h	min
	3	10		4	20		6	30		5	35
	−1	50		−2	30		−4	40		−3	55

9	h	min	**10**	h	min	**11**	h	min	**12**	h	min
	6	20		3	15		9	35		4	20
	−1	50		−1	55		−3	45		−3	35

13	min	sec	**14**	min	sec	**15**	min	sec	**16**	min	sec
	5	10		10	20		7	20		8	40
	−2	10		−1	10		−4	35		−5	50

17	days	h	**18**	days	h	**19**	days	h	**20**	days	h
	6	20		9	14		4	15		12	17
	−2	15		−3	12		−1	9		−5	5

21	days	h	**22**	days	h	**23**	days	h	**24**	days	h
	6	5		8	3		4	5		6	15
	−2	18		−4	17			−21		−3	23

25	days	h	**26**	days	h	**27**	days	h	**28**	days	h
	7	12		8	10		9	3		11	2
	−6	15		−3	19		−2	14		−4	12

TIME SUBTRACTION

Now try these

1	h	min	**2**	h	min	**3**	h	min	**4**	h	min
	5	30		3	45		4	38		3	47
	− 3	15		− 1	20		− 2	16		− 2	25
5	h	min	**6**	h	min	**7**	h	min	**8**	h	min
	15	10		24	40		13	35		34	10
	− 9	30		− 16	45		− 1	45		− 8	50
9	h	min	**10**	h	min	**11**	h	min	**12**	h	min
	25	8		33	14		17	26		35	20
	− 1	32		− 28	45		− 12	39		− 21	31
13	min	sec	**14**	min	sec	**15**	min	sec	**16**	min	sec
	15	50		40	45		27	18		54	12
	− 10	13		− 30	30		− 11	46		− 39	54
17	days	h	**18**	days	h	**19**	days	h	**20**	days	h
	9	15		14	20		25	18		36	10
	− 6	10		− 7	16		− 18	15		− 6	5
21	days	h	**22**	days	h	**23**	days	h	**24**	days	h
	45	0		35	2		46	8		38	3
	− 2	12		− 3	15		− 28	22		− 20	20
25	days	h	**26**	days	h	**27**	days	h	**28**	days	h
	56	17		49	11		70	0		81	8
	− 55	19		− 6	13		− 53	16		− 9	14

TIME MULTIPLICATION AND DIVISION

First try these

Set 1

Multiply the following quantities by 2

1	3 h 15 min	**2**	4 h 20 min
3	1 h 30 min	**4**	2 h 40 min
5	3 h 21 min	**6**	5 h 35 min
7	4 min 25 sec	**8**	1 min 30 sec
9	6 min 35 sec	**10**	2 min 50 sec
11	3 days 7 h	**12**	4 days 5 h
13	1 day 11 h	**14**	5 days 6 h
15	5 days 12 h	**16**	8 days 14 h
17	7 days 10 h	**18**	6 days 17 h

Set 2

Divide the following quantities by 2

1	6 h 0 min	**2**	10 h 20 min
3	5 h 0 min	**4**	3 h 10 min
5	8 h 40 min	**6**	7 h 12 min
7	4 min 0 sec	**8**	6 min 30 sec
9	12 min 2 sec	**10**	11 min 4 sec
11	4 days 12 h	**12**	10 days 18 h
13	6 days 22 h	**14**	9 days 0 h
15	3 days 4 h	**16**	7 days 2 h
17	12 days 9 h	**18**	5 days 6 h

TIME MULTIPLICATION AND DIVISION

Now try these

Set 1

Multiply the following quantities by 2

1	1 h 25 min	**2**	6 h 30 min
3	5 h 45 min	**4**	7 h 52 min
5	9 h 23 min	**6**	11 h 39 min
7	8 min 29 sec	**8**	17 min 31 sec
9	13 min 48 sec	**10**	15 min 18 sec
11	6 days 10 h	**12**	9 days 12 h
13	14 days 15 h	**14**	23 days 19 h
15	18 days 11 h	**16**	10 days 21 h
17	16 days $13\frac{1}{2}$ h	**18**	25 days 23 h

Set 2

Divide the following quantities by 2

1	8 h 30 min	**2**	11 h 0 min
3	9 h 2 min	**4**	17 h 24 min
5	21 h 50 min	**6**	16 h 58 min
7	12 min 46 sec	**8**	15 min 14 sec
9	29 min 56 sec	**10**	38 min 52 sec
11	8 days 14 h	**12**	18 days 20 h
13	19 days 8 h	**14**	35 days 17 h
15	54 days 21 h	**16**	43 days 12 h
17	37 days 15 h	**18**	61 days 23 h

TIME PROBLEMS

First try these

1 Write in figures:

(*a*) Twenty minutes past 15:00
(*b*) Twenty-five minutes to 10:00
(*c*) Six minutes past 09:00
(*d*) Three minutes to 17:00
(*e*) Twenty-nine minutes past noon
(*f*) Fourteen minutes to 20:00

2 How many hours and minutes are there between:

(*a*) 06:25 and 09:10 (*b*) 15:01 and 19:58
(*c*) 01:28 and 02:09 (*d*) 21:16 and 23:43
(*e*) 10:50 and 12:20 (*f*) 23:13 and 01:26

3 How many days and hours are there between:

(*a*) 09:00 on Monday and 10:00 on the following Tuesday?
(*b*) 18:00 on Saturday and 04:00 on the following Monday?
(*c*) 14:00 on Friday and 23:00 the next day?
(*d*) 08:00 today and 17:00 yesterday?
(*e*) Noon today and midnight tomorrow?

4 A man drives 8 km in 9 min. If he drives at the same speed, how long would he take to drive 40 km?

5 A man is paid at the rate of £1·20 per hour. He received £6·00 for his work. How many hours did he work?

6 Find the sum of 3 h 25 min and 6 h 49 min, and subtract the result from 24 h.

7 A girl started on a journey at 09:10 and arrived at 13:25. How long was she on the journey?

8 It takes a bus 7 h 43 min each way on a journey. How long does it take to complete the return journey, allowing a stop of 1 h before beginning the return trip?

9 A ship was due at noon on Monday, but arrived at 15:00 on Tuesday. How late was it?

TIME PROBLEMS

Now try these

1 Write in figures:
   (*a*) Thirty-seven minutes past 01:00
   (*b*) Eighteen minutes to midnight
   (*c*) Twenty-nine minutes to 04:00
   (*d*) One minute to noon

2 What is the difference in time between:
   (*a*) 12 h 6 min 14 sec and 12 h 8 min 3 sec
   (*b*) 4 h 20 min 36 sec and 4 h 24 min 30 sec
   (*c*) 0 h 58 min 47 sec and 1 h 1 min 13 sec

3 How many hours and minutes are there between:
   (*a*) 04:37 and 09:15
   (*b*) 06:05 and 11:42
   (*c*) 08:40 and 14:29
   (*d*) 10:19 and 17:38
   (*e*) 19:26 and 01:09
   (*f*) 15:15 and 09:17

4 How many days and hours are there between:
   (*a*) 06:00 on Sunday and 18:00 on the following Thursday?
   (*b*) 03:30 on Friday and 16:30 on the Tuesday following?
   (*c*) 12:20 on Saturday and 19:20 on the day after next?

5 If the clocks in Sydney, Australia, are $9\frac{1}{2}$ h in front of London clocks, what time is it in Sydney when it is 08:42 in London?

6 A swimmer took 16 h 32 min in swimming the English Channel. If he entered the water on the French side at 01:25, at what time did he arrive at the English coast?

7 A cyclist leaves home at 08:35 and travels 88 km at a steady speed of 16 km per hour. At what time does he finish his journey?

8 A bus driver goes on duty at 06:00 and completes 11 journeys, each taking 43 min. At the end of the 5th journey he has half-an-hour break. At what time does he leave work?

DIRECTION

All can try these

1 The direction opposite to N is ....

2 The direction opposite to E is ....

3 The direction opposite to SE is ....

4 The direction opposite to NE is ....

5 If you face N and then right turn, in which direction will you be facing?

6 If you face W and then left turn, in which direction will you be facing?

7 If you face NW and then about turn, in which direction will you be facing now?

8 A part of a river runs directly east and west. What would be the direction of a bridge across the river at this point?

9 The windows of our classroom look southwards, and the light from these windows comes in on our left. In which direction do we face when we sit in class?

10 A girl walks directly westwards along a street and then turns sharply to the right. In which direction is she walking now?

11 A boy walks due south along a street and then turns sharply to the left. In which direction is he walking now?

12 A boy walks round a square. He starts from one corner walking towards the north. In which direction will he be walking when he returns to his starting-point?

13 SW is opposite to ....

14 NW is opposite to ....

15 A boy is walking along a road in an easterly direction. He does an about turn. In what direction is he then walking?

16 A playground is in the shape of a square. A boy leaves the north-east corner and walks along the diagonal to the opposite corner. In what direction is he walking, and on which side of him is south?

ANGLES

All can try these

1 Fold a piece of paper carefully in two, and then into two again. The corner you have just made is called a square corner, or a right angle.

2 Lay your pencil along one side of the square corner and turn it until it lies along the other side. You have turned your pencil through a right angle.

3 A right angle is divided up into much smaller angles called degrees, written °. There are ninety degrees in a right angle, written 90°. Write sixty degrees in the same way.

4 Write forty degrees and seventy degrees.

5 How many degrees are there in half a right angle?

6 If I turn from facing west to facing north, through how many degrees do I turn?

7 How many degrees are there in two right angles?

8 Lay your pencil on the table and turn it through two right angles. Through what shape have you turned it?

9 Lay your pencil on the table and turn it through four right angles. Through what shape have you turned it?

10 As the hands of a clock turn, they make angles with each other. At what angle are the hands of a clock at 0900 h and 1800 h?

11 At what angle are the hands of a clock at 1500 h?

12 At what angle are the hands of a clock at 1200 h?

13 At what angle are the hands of a clock at 1630 h?

14 At what angle are the hands of a clock at 1930 h?

15 ABC is a straight line. If angle DBC (marked) is 37°, how many degrees are there in angle DBA?

D

A B C

16 How many degrees must be added to 124° to make three right angles?

SHAPES

All can try these

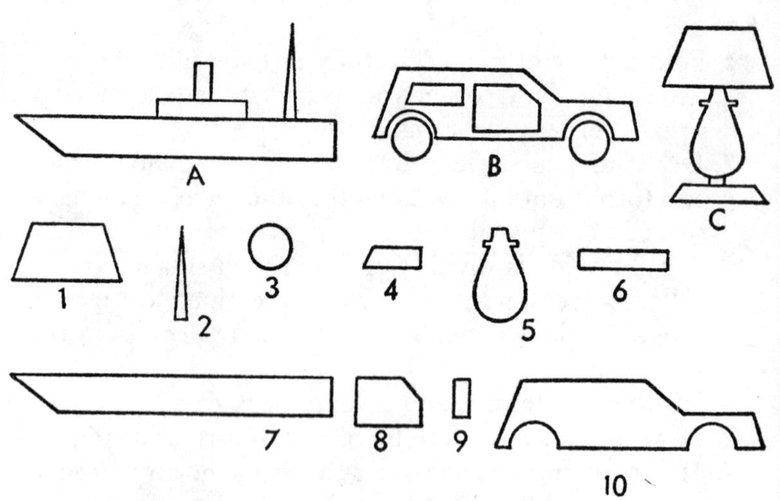

1 (*a*) Drawing 1 is a part of which picture?
(*b*) Drawing 2 is a part of which picture?
(*c*) Drawing 3 is a part of which picture?
(*d*) Drawing 4 is a part of which picture?
(*e*) Drawing 5 is a part of which picture?
(*f*) Drawing 6 is a part of which picture?
(*g*) Drawing 7 is a part of which picture?
(*h*) Drawing 8 is a part of which picture?
(*i*) Drawing 9 is a part of which picture?
(*j*) Drawing 10 is a part of which picture?

2 Print the word **LUNCHES** as it would appear in a mirror.

3 Which capital letters of the alphabet would look exactly the same in a mirror?

4 The shape of a butterfly could be folded so that the two halves correspond exactly. Such a shape is called symmetrical. What other objects can you think of whose shapes are symmetrical?

SHAPES

All can try these

Look at the shapes below. Imagine that you are looking at the shapes in a mirror. Draw the reflection that you would see. The first one is done for you.

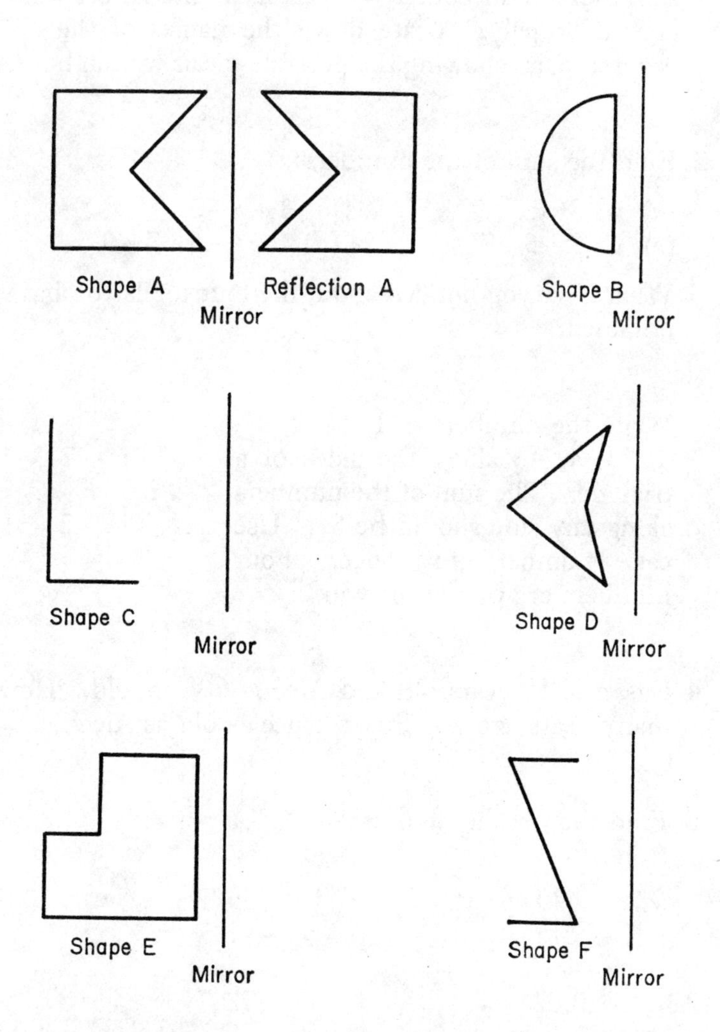

FINDING OUT 1

All can try these

1 For our first course at lunch we can have steak or fish. For the second course we can have cherry pie, ice-cream, or jelly. Write down the names of the six ordered pairs showing the possible meals we can have.

2 Find the sum of the numbers:

(*a*) $1 + 3$ (*b*) $1 + 3 + 5$
(*c*) $1 + 3 + 5 + 7$ (*d*) $1 + 3 + 5 + 7 + 9$

What have you noticed about the four totals you have obtained?

3 Write the numbers $\frac{1}{2}$, 1, $1\frac{1}{2}$, 2, $2\frac{1}{2}$, 3, $3\frac{1}{2}$, 4, $4\frac{1}{2}$ along the sides of a triangle. The sum of the numbers along any side should be $8\frac{1}{2}$. Use each number only once. Four numbers are placed for you.

```
         1/2
       ?     ?
     ?         ?
 1 1/2   ?   2    1
```

4 Susan is 10 years old and Alec is 6 years old. How many years ago was Susan twice as old as Alec?

5 Find the missing figures:

(*a*)
```
   ?49
 + 6??
 -----
  ?007
```

(*b*)
```
  30??
 - ?05
 -----
  ?207
```

FINDING OUT 2

All can try these

1 What number can you add to or subtract from 5 and get the same answer?

2 What is the temperature in your classroom?

3 What kind of number do you always get when you add two even numbers?

4 What number must I multiply 36 by to get 252?

5 How many days are there in a leap year?

6 How much does the figure 8 stand for in each of the numbers 834, 286, 178?

7 Through what angle does the large hand of a watch turn in moving from quarter past to half past the hour?

8 Is a square always a rectangle?

9 With what figure does the answer end when you subtract 7 from a number ending in 3?

10 If a means 1 and b means 10 we can write 36 as $3b + 6a$. Write 29, 55, 86, 97 and 108 in terms of a and b.

11 Use the figures 4, 5, 6, 7 to write (*a*) the largest four-place number possible and (*b*) the smallest four-place number possible.

12 Arrange those numbers in order of size beginning with the smallest number 1101, 1110, 1011.

13 What unit of measure would be used to measure (*a*) the temperature inside a refrigerator, (*b*) the weight of a pencil, (*c*) the length of a board, (*d*) the quantity of milk?

14 What kind of number do you always get when you add two odd numbers?

COMPARING MEASURES

All can try these

Write 'Yes' if the sentence is reasonable. Write 'No' if it is wrong or not reasonable.

1 Some months have 5 Mondays.

2 A dog can run at a speed of 30 km an hour.

3 A car is more than 30 m long.

4 An elephant weighs more than 500 kg.

5 A boy should drink 8 litres of milk each day.

6 Some years have more than 12 months.

7 There are more than 100 cups of milk in 4 litres of milk.

8 A pencil weighs about 60 g.

9 Some tall buildings in big cities are about a kilometre high.

10 A rectangle can be larger than a square.

11 A dozen pencils are fewer than a score of pencils.

12 A girl aged 9 years weighs about 5 kg.

13 A metre is longer than 110 cm.

14 A car will travel 10 km on a litre of petrol.

15 A train can travel at 110 km an hour.

16 A kg of iron weighs more than 1 kg of feathers.

17 The number 7077 is larger than the number 7707.

18 A motor cycle could travel 200 km on 1 litre of petrol.

19 A letter weighs 30 g.

20 The usual size of a milk bottle is rather more than one half-litre.

COMPLETING SETS OF NUMBERS

All can try these

Set 1

Put the two missing numbers into each of these sets:

1 1, 2, 3, 4 — —

2 3, 5, 7, 9 — —

3 8, 11, 14, 17 — —

4 20, 18, 16, 14 — —

5 35, 30, 25, 20 — —

6 49, 42, 35, 28 — —

7 2—1, 4—2, 6—3, 8—4 — —

8 6, 12, 18, 24 — —

9 108, 99, 90, 81 — —

10 1, 2, 4, 7, 11 — —

Set 2

1 1, 2, 4, 8 — —

2 3, 6, 12, 24 — —

3 5, 15, 45, 135 — —

4 $\frac{1}{2}$, 2, 8, 32 — —

5 1, 5, 25, 125 — —

6 96, 48, 24, 12 — —

7 1, 3, 9, 27 — —

8 2, 4, 8, 16 — —

9 1024, 256, 64, 16 — —

10 144, 72, 36, 18 — —

GENERAL REVISION

First try these

Add

1

litres	ml
14	270
6	59
+10	815

2

m	cm
7	28
3	12
+1	79

3

£
15·68
$6{\cdot}97\frac{1}{2}$
$+0{\cdot}08\frac{1}{2}$

Subtract

4

h	min
15	21
− 9	48

5

kg	g
82	15
−30	600

6

1207
− 498

Multiply

7 2 m 48 cm × 2
8 7 min 25 sec × 2
9 67 × 45

Divide

10 114 litres ÷ 6
11 875 ÷ 21
12 13 km 500 m ÷ 2

13 Write in figures one thousand five hundred and forty.
14 Find the total cost of 5 pairs of stockings at 50p per pair, and 3 pairs at 75p per pair.
15 How many minutes are there between midnight and 01:47?
16 Take 347 from 906, and divide the result by 13.
17 If 40 litres of water is allowed for one man per day, how much water should he be allowed for a week of 7 days?
18 A man earns £1·05 per hour. How much will he earn in (*a*) a day of 8 working hours, and (*b*) a week of 5 days?
19 Find the sum of 8 min 3 sec, 21 min 50 sec, 9 min 42 sec and 19 sec.
20 Take 75 g from $4\frac{1}{2}$ kg, and add 15 g to your result.
21 How many grammes are there in (*a*) 4 kg, (*b*) $6\frac{1}{2}$ kg?
22 What must be added to $2\frac{1}{2}$ m to make 5 m 10 cm?

GENERAL REVISION

Now try these

Add

1

kg	g
74	105
28	33
+ 1	426

2

km	m
18	621
9	709
+56	284

3

days	h
12	19
6	12
+ 9	15

Subtract

4

litres	ml
46	0
− 19	760

5

h	min
35	29
− 14	54

6

m	cm
96	31
− 25	74

Multiply

7 93 × 58

8 10 min 41 sec × 2

9 16 litres × 11

Divide

10 37 kg 300 g ÷ 2

11 59 litres 750 ml ÷ 2

12 1147 ÷ 37

13 Write in figures six thousand nine hundred and seven.

14 A box holds 25 kg of sugar. What weight of sugar is in 10 boxes? How much less than 300 kg is this weight?

15 From a tank holding 120 litres of oil a shopkeeper sold 63 litres and 50 half-litres. How much oil was left?

16 Find the total cost of a table costing £70·00, 4 chairs priced £15·15 each and a sideboard costing £75·50.

17 Divide 4 days 12 h into 3 equal periods of time.

18 If $\frac{1}{2}$ litre of water weighs $\frac{1}{2}$ kg, what is the weight of $10\frac{1}{2}$ litres of water?

19 Add together 53, 363, 197, 209 and 425, and divide the result by 29.

20 Find the difference between 1000 m and $721\frac{1}{2}$ m.

21 Add together 20 kg, $4\frac{1}{2}$ kg and $40\frac{1}{2}$ kg, and subtract the result from 100 kg.

22 A certain job takes 4 days 13 h. How many days and hours will it take to repeat the job 4 times?

USE OF SYMBOLS

All can try these

If ○ + ○ = □, and □ + □ = △:

(*a*) How many times more than ○ is □ worth?

(*b*) How many times more than ○ is △ worth?

(*c*) How many ○'s is □ + □ worth?

(*d*) How many ○'s is △ + △ worth?

(*e*) How many ○'s is □ + △ worth?

(*f*) How many ○'s is △ − □ worth?

(*g*) How many ○'s is □ + □ + △ worth?

Now work these exercises. Use as few symbols as possible in your answer and put your answer in terms of the largest symbols possible. That is to say, do not write ○ + ○ or 2○ in your answer but put □, and do not write □ + □ or 2□ but put △. The first two exercises are done for you.

Set 1

(*i*) □ + ○ + ○ = △ (*ii*) ○ + ○ + □ + □ = △ + □

(*iii*) □ + □ + △ = (*iv*) ○ + ○ + △ =

(*v*) □ + △ − ○ = (*vi*) ○ + ○ + ○ + ○ + □ + □ =

(*vii*) □ + □ − ○ − ○ = (*viii*) △ − □ − ○ =

(*ix*) ○ + △ − □ = (*x*) □ + ○ + ○ + △ =

Set 2

(*i*) ○ + ○ + △ + △ + □ =

(*ii*) △ + △ + ○ + ○ + ○ + ○ + □ =

(*iii*) ○ + △ + ○ + ○ + □ − △ =

(*iv*) □ + □ + ○ + ○ + △ − △ − △ =

(*v*) ○ + ○ + □ + △ + ○ − ○ =